Plant Biology

Series Editor
Nina Strömgren Allen
Wake Forest University
Department of Biology
Winston-Salem, North Carolina

Advisory Board
Lawrence Bogorad
Harvard University
Cambridge, Massachusetts

Joseph E. Varner
Washington University
St. Louis, Missouri

R. Malcolm Brown, Jr.
University of Texas
Austin, Texas

Milton Zaitlin
Cornell University
Ithaca, New York

Maureen R. Hanson
Cornell University
Ithaca, New York

Titles in the Series

Algae as Experimental Systems

FRONTISPIECE: "Algaprint," made by photographing an agar plate containing the cyanophyte *Phormidium* after exposing the plate to the light passing through a negative of a photograph of the Freiburg cathedral.

Phormidium cells move by gliding. At a light-dark threshold, cells turn counter-clockwise toward the light, thus accumulating in the lighted portions of the field. Sensitivity is sufficient to respond differentially to a 4% difference in light intensity. *Photograph courtesy of Donat-Peter Hader.*

Algae as Experimental Systems

Editors

Annette W. Coleman
Division of Biology and Medicine
Brown University
Providence, Rhode Island

Lynda J. Goff
Department of Biology
University of California, Santa Cruz
California

Janet R. Stein-Taylor
Department of Biological Sciences
University of Illinois at Chicago
Illinois

Alan R. Liss, Inc., New York

Randall Library UNC-W

Address all Inquiries to the Publisher
Alan R. Liss, Inc., 41 East 11th Street, New York, NY 10003

Copyright © 1989 Alan R. Liss, Inc.

Printed in the United States of America

Under the conditions stated below the owner of copyright for this book hereby grants permission to users to make photocopy reproductions of any part or all of its contents for personal or internal organizational use, or for personal or internal use of specific clients. This consent is given on the condition that the copier pay the stated per-copy fee through the Copyright Clearance Center, Incorporated, 27 Congress Street, Salem, MA 01970, as listed in the most current issue of "Permissions to Photocopy" (Publisher's Fee List, distributed by CCC, Inc.), for copying beyond that permitted by sections 107 or 108 of the US Copyright Law. This consent does not extend to other kinds of copying, such as copying for general distribution, for advertising or promotional purposes, for creating new collective works, or for resale.

Library of Congress Cataloging-in-Public Data

Meeting on Algal Experimental Systems in Cell
   Biological Research (1988 : Airlie, Va.)
   Algae as experimental systems.

   (Plant biology ; v. 7)
   Includes bibliographies and indexes.
   1. Plant cells and tissues—Congresses.
2. Algae—Cytology—Congresses. I. Coleman,
Annette W. II. Goff, Lynda J. III. Stein-Taylor,
Janet R. IV. Title. V. Series.
QK725.M44 1988   581'.07'24   88-37733
ISBN 0-8451-1806-4

We thank the publishers for permission to reproduce the following figures.
The Journal of Cell Biology by copyright permission of the Rockefeller University Press: Figures 1,2,3,4,6, McDonald and Cande; Figures 2,3,6, Salisbury; Figure 1, Gantt.
Nature copyright 1985 Macmillan Journals Limited: Figure 5, McDonald and Cande.
Kluwer Academic Publishers: Figure 2, Grossman et al.

QK
725
.A43
1989

Contents

Section III.   Cellular Recognition

Section IV.   Model Systems for Photoreactions and Rhythms

Contributors

W. Steven Adair, Department of Anatomy and Cell Biology, Tufts Medical School, Boston, MA 02111 **[171]**

Nina S. Allen, Wake Forest University, Department of Biology, Winston-Salem, NC 27109 **[xi]**

Lamont K. Anderson, Carnegie Institution of Washington, Department of Plant Biology, Stanford, CA 94305 **[269]**

Peter L. Beech, School of Botany, University of Melbourne, Parkville 3052, Victoria, Australia **[93]**

W. Zacheus Cande, Department of Botany, University of California, Berkeley, CA 94720 **[3]**

Annette W. Coleman, Division of Biology and Medicine, Brown University, Providence, RI 02912 **[xiii]**

P.B. Conley, V.A. Medical Center M-151, Palo Alto, CA 94304 **[269]**

R. Demets, Department of Molecular Cell Biology, University of Amsterdam, The Netherlands **[187]**

Susan K. Dutcher, Department of Molecular, Cellular, and Developmental Biology, University of Colorado, Boulder, CO 80309-0347 **[39]**

Christine Elsner-Menzel, Max-Planck-Institut für Zellbiologie, D-6802 Ladenburg, Federal Republic of Germany **[71]**

Kenneth W. Foster, Department of Physics, Syracuse University, Syracuse, NY 13244-1130 **[215]**

Elisabeth Gantt, Smithsonian/Botany Department, University of Maryland, College Park, MD 20742 **[249]**

Lynda J. Goff, Department of Biology, University of California, Santa Cruz, CA 95064 **[xiii]**

Ursula W. Goodenough, Biology Department, Washington University, St. Louis, MO 63130 **[171]**

Arthur R. Grossman, Carnegie Institution of Washington, Department of Plant Biology, Stanford, CA 94305 **[269]**

Lothar Jaenicke, Institut für Biochemie, D-5000 Köln 1, Federal Republic of Germany **[135]**

Oswald Kiermayer, Institute of Plant Physiology, University of Salzburg, A-5020 Salzburg, Austria **[149]**

R. Kooijman, Department of Molecular Cell Biology, University of Amsterdam, The Netherlands **[187]**

The numbers in brackets are the opening page numbers of the contributors' articles.

Anthony Koutoulis, School of Botany, University of Melbourne, Parkville 3052, Victoria, Australia **[93]**

Darryl L. Kropf, Department of Botany/Plant Pathology, Oregon State University, Corvallis, OR 97331; present address; Department of Biology, University of Utah, Salt Lake City, UT 84112 **[111]**

John W. La Claire, Department of Botany, University of Texas at Austin, Austin, TX 78713-7640 **[55]**

P.G. Lemaux, Pfizer Inc., Central Research, Groton, CT 06340 **[269]**

Kent McDonald, Department of Molecular, Cellular and Developmental Biology, University of Colorado, Boulder, CO 80309 **[3]**

Ursula Meindl, Institute of Plant Physiology, University of Salzburg, A-5020 Salzburg, Austria **[149]**

Diedrik Menzel, Max-Planck-Institut für Zellbiologie, D-6802 Ladenburg, Federal Republic of Germany **[71]**

Kenneth R. Miller, Division of Biology and Medicine, Brown University, Providence, RI 02912 **[233]**

Dieter G. Müller, Fakultät für Biologie der Universität, D-7750 Konstanz, Federal Republic of Germany **[201]**

Ralph S. Quatrano, Department of Botany/Plant Pathology, Oregon State University, Corvallis, OR 97331; present address: Department of Biology, University of North Carolina, Chapel Hill, NC 27599-3280 **[111]**

Jeffrey L. Salisbury, Center for NeuroSciences, Case Western Reserve University, School of Medicine, Cleveland, OH 44106 **[19]**

Jureepan Saranak, Department of Physics, Syracuse University, Syracuse, NY 13244-1130 **[215]**

Laurel Spear-Bernstein, Department of Molecular Biology, Scripps Clinic and Research Foundation, La Jolla, CA 92037 **[233]**

Richard C. Starr, Department of Botany, The University of Texas at Austin, Austin, TX 78713 **[135]**

Beatrice M. Sweeney, Department of Biological Sciences, University of California, Santa Barbara, CA 93106 **[289]**

A.M. Tomson, Department of Molecular Cell Biology, University of Amsterdam, The Netherlands **[187]**

H. van den Ende, Department of Molecular Cell Biology, University of Amsterdam, The Netherlands **[187]**

Susan D. Waaland, Department of Biology, University of Puget Sound, Tacoma, WA 98416 **[121]**

Richard Wetherbee, School of Botany, University of Melbourne, Parkville 3052, Victoria, Australia **[93]**

Foreword

Algae provide a rich and varied source of biological material to study many diverse natural processes. A cursory inspection of the composition of the chapters contained in this, the seventh volume in the PLANT BIOLOGY series, can confirm for the curious reader the wealth of information waiting to be gleaned from algae as experimental systems.

This editorial undertaking was originally planned by a small, focused group of phycologists who gathered at my suggestion during the North East Algal Symposium held in Woods Hole, Massachusetts in the Spring of 1985. At that meeting, I asked Annette W. Coleman and Lynda J. Goff to consider editing for the PLANT BIOLOGY series a volume that would include, within the limitations of space, the many and diverse modern approaches to the cellular and molecular biology of algae. They enthusiastically agreed to undertake this task and also developed a scientific conference on the same topic. They provided inspired, energetic and expert guidance throughout these projects.

It is my hope that this book will both inspire scientists already working with algae to further their efforts, as well as provoke others not yet enlightened about algae to use them as systems to address many important questions and challenges in biology.

I want to thank Janet Stein-Taylor for her superb job in directing the prompt review and careful editing of the volume. I thank also Hannah Croasdale of Dartmouth College for helping me truly understand how wonderful and exciting it is to study all facets of algae. The energy, care and inspiration Hannah Croasdale has brought to the study of algae over the last 60 or so years should be an inspiration to us all.

Nina Strömgren Allen
Series Editor

Preface

The organisms known as the algae represent a myriad of evolutionary lineages and experiments in adaptation. Their ancestral forms have given rise to both plants and animals. Hence, nearly all the structures and functions that have evolved in eukaryotic organisms can be studied in one or more of the many groups of algae. The result is that valuable and unexpected insights into basic functions have been revealed through comparative studies of their biology.

The use of algae for landmark discoveries in basic biology has a long history. The most notable examples include 1) the discovery of the classic "9 + 2" arrangement of the flagellar axoneme, which was first recognized in an alga (Manton, 1952); 2) the undeniable existence of long-lived messenger RNA, which was elegantly shown in enucleated and grafted *Acetabularia* (Hämmerling, 1953); 3) the first definitive action spectrum for photosynthesis, which Engelmann (1882) ingeniously demonstrated by the orientation of bacteria alongside the green alga *Cladophora* irradiated with light from a prism; and 4) the long-sought first products of carbon dioxide fixation, which were isolated finally from the unicellular green alga *Chlorella* (Calvin and Benson, 1948). In addition, the elegant but flawed research of Moewus on *Chlamydomonas* sexuality (see Gowans, 1976) nonetheless served the invaluable function of focussing research on this exemplary organism. Following Sager's (1954) discovery of its uniparental inheritance of plastid characters, the use of *Chlamydomonas* as a model system for basic research has expanded markedly.

All of these discoveries result from two characteristics shared by algae, both of which are of enormous benefit to researchers: most algal cells are easy to handle in the laboratory and they are relatively small in size and simple in construction, which permits direct observation of the effects of experimental manipulation on intact cells, a valuable supplement to physiological, biochemical, and genetic analyses. Furthermore, pure pedigreed cultures are available from national culture collections. Nevertheless, much remains to be discovered, and algae as research subjects will continue to have vital roles in major discoveries in biology.

In June 1988 the American Society for Cell Biology sponsored and generously supported a summer conference on "Algal Experimental Systems in Cell Biological Research." It brought together 121 active researchers from ten countries to present papers and posters. This meeting provided an unrivalled opportunity for workers on disparate problems in cell biology, developmental biology, and molecular biology to share recent results and methodologies. Major papers were presented on the structure, genetics, and behavior of flagella; the control of cytoskeletal elements and of cell wall construction; the origin of cell polarity; activities essential to cell division; interactions in cell recognition; the use of mutants to study differentiation; the basis of circadian rhythms; and the elements of the light-harvesting machinery of photosynthesis. Sixty-four posters on a wide range of topics were displayed and discussed.

Expanded, refereed manuscripts of the major presentations, designed both to review past research and to define major questions and future directions, are included in this volume, along with a listing of poster presentations. This work provides a synopsis of the history and current status of research on numerous problems of basic cell biology. It also presents a broad sampling of current research with algal cells and the variety of organisms being used.

For those who teach, for those who learn, and for those whose research would profit from the use of an algal model system, we hope this volume will serve as a source of reference and inspiration, creating an awareness of the potentials and advantages of the algae as research tools.

REFERENCES

Calvin M, Benson AA (1948). The path of carbon in photosynthesis. Science 107:476–480.

Engelmann TW (1882). On the production of oxygen by plant cells in a microspectrum. Bot Zeit 40:419–426.

Gowans CS (1976). Publications by Franz Moewus on the genetics of algae. In Lewin RA (ed): "The Genetics of Algae." Berkeley: University of California Press, pp 310–332.

Hämmerling J (1953). Nucleocytoplasmic interactions in *Acetabularia* and other cells. Annu Rev Plant Physiol 14:65–92.

Manton I (1952). The fine structure of plant cilia. Symp Soc Exp Biol 6:306–319.

Sager R (1954). Mendelian and non-mendelian inheritance of streptomycin resistance in *Chlamydomonas reinhardtii*. Proc Natl Acad Sci USA 40:356–363.

Annette W. Coleman
Lynda J. Goff

Section I.
Cytoskeleton Components

Algae as Experimental Systems pages 3–18
© 1989 Alan R. Liss, Inc.

DIATOMS AND THE MECHANISM OF ANAPHASE SPINDLE ELONGATION

Kent McDonald and W. Zacheus Cande

Department of Molecular, Cellular and
Developmental Biology, University of
Colorado, Boulder, Colorado 80309 (K.M.);
Department of Botany, University of
California, Berkeley, California 94720 (W.Z.C.)

Until recently, algae have not been much utilized as
experimental model systems to study mitosis, as confirmed by
reading the literature of mitosis prior to ca. 1976
(Schrader, 1953; Nicklas, 1971; Luykx, 1970; Bajer and Mole-
Bajer, 1972; Inoue and Stephens, 1975) where the only
reference is to Manton et al. (1969a, b, 1970a, b) on the
diatom Lithodesmium. The considerable amount of work on
algal mitotic ultrastructure (Pickett-Heaps, 1974; Heath,
1980) was primarily descriptive rather than experimental and
the results were used mainly to evaluate phylogenetic
relationships as opposed to the mechanisms of mitosis per
se. After 1975, diatoms began to be used as serious model
systems for the study of spindle elongation, first by
electron microscopy (McDonald et al., 1977; Pickett-Heaps &
Tippit, 1978), as subjects for UV microbeam experiments
(Leslie & Pickett-Heaps, 1983, 1984), and then as systems
which could be reactivated in vitro (Cande & McDonald, 1985;
Cande, 1986; Cande et al., 1988). The chief reason for the
appeal of diatom spindles is the nearly paracrystalline
arrangement of spindle microtubules associated with spindle
elongation.

The importance of structural regularity cannot be
overemphasized when it comes to understanding motility
phenomena. The main reason we know as much as we do about
striated muscle and ciliary motility is their high degree of
order coupled with the ability to use cell models for making
structural-functional correlations. It follows that our best
hope for understanding mitosis is with spindles which are
the most highly ordered, and at our current state of

knowledge this certainly includes diatom spindles. In the text that follows, we will first consider the mitosis problem in general, then review the details of diatom spindle ultrastructure. Finally, we will see how the Stephanopyxis turris in vitro model system has contributed to our understanding of the mechanism of anaphase spindle elongation.

MITOSIS AS A RESEARCH PROBLEM

Mitosis has been studied seriously as a research problem for well over a century - so why don't we understand more about mitotic mechanisms than we do? The answer is related not only to the complexity of the process itself but also to the tools used to study the problem. For most of this century-long investigation the primary tool has been the light microscope. While this is a necessary tool in mitosis research, it is not sufficient. Mitosis is a problem which must be described and modelled at the molecular level. Thus, we need a detailed description of the mitotic apparatus at electron microscope or X-ray crystallography levels of resolution. We need biochemical descriptions of the structural, mechanochemical and regulatory molecules involved in moving chromosomes. We need to know the relationship between the structures seen in the electron microscope and the molecules seen as bands on a gel. Finally, we need a way to check the structure-function relationships implied or deduced from the ultrastructural and biochemical information. When we consider the problem in these terms, it is not so surprising that we don't understand very much about the way chromosomes move. What it amounts to is that we have a lot of information which is not particularly useful by itself (light microscope descriptions) and very little information that we really need (descriptions at the molecular level). And the molecular level information that we do have is made even less useful because it comes from widely diverse sources.

Chromomosomes are moved at least five different ways during mitosis: 1) nuclear envelope-associated prophase movements; 2) prometaphase congression; 3) anaphase A (movement toward the poles); 4) anaphase B (spindle elongation); and, 5) nuclear/chromosomal migrations in telophase or pre-prophase. It is unlikely that each of these movements has a totally different mechanochemistry but we cannot say for sure that they do not. What we can say is that failure to recognize the diversity of chromosome

movements during mitosis can lead to erroneous interpretation of experimental results or models which are too
simplistic.

One way in which the mitotic spindle is different from
most other motility systems is its lability. It forms at
prophase, functions during anaphase to separate the
chromosomes, then disappears at telophase. This means that
the factors which regulate the assembly and disassembly of
microtubules (MTs) and the structure of MTs themselves
become important to our understanding of mitotic mechanisms.
MTs have an inherent molecular polarity due to the helical
arrangement of tubulin subunits. Biochemical and
immunofluorescence studies (see Cassimeris et al., 1987)
have shown that one end of a microtubule is the preferred
end for addition and removal of subunits. This is referred
to as the plus end. The other end (the minus end) will also
exchange subunits but at slower rates. Using modified
tubulin (Heidemann & McIntosh, 1980) or flagellar dynein
molecules (Telzer & Haimo, 1981) it is possible to identify
the molecular polarity of MTs with the electron microscope.
Studies by Euteneuer and McIntosh (1980, 1981) show that
mitotic spindles have all their MTs arranged such that the
minus ends are at the poles and the plus ends either at
kinetochores (kinetochore MTs), free in the spindle (polar
MTs), or closely aligned with MTs from the opposite pole
(interdigitating MTs). MTs with both ends free (free MTs) or
which extend completely from pole to pole (continuous MTs)
are rare in most spindles. Although diatoms have not been
studied with these polarity markers we can assume that they
follow the conventions established for all other spindles
studied. As far as we can tell from the limited data
describing the different kinds of MTs and their distribution
in other spindle types, diatoms are unusual only in the
sense that they have all their interdigitating MTs arranged
in a paracrystalline array rather than distributed
throughout the spindle in small bundles as in other
spindles, e.g., in PtK1 (McDonald & Euteneuer, 1983).

STRUCTURAL BACKGROUND

A discussion of algal spindle structure studies and in
particular diatom spindles must begin with the remarkable
work of Robert Lauterborn (1896). He described the
formation and behavior of centrosomes and mitotic spindles

Figure 1. Longitudinal section of a Stephanopyxis turris central spindle at prometaphase: many parallel MTs insert into each spindle pole body (P) and overlap in the midregion which appears darker than the rest of the spindle. Bar = 0.5 um. (From McDonald et al., 1986).

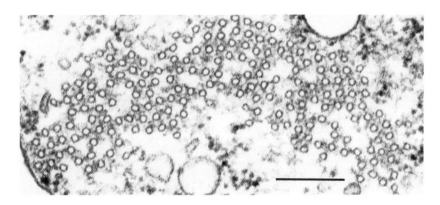

Figure 2. Cross section through the overlap region of a Diatoma vulgare central spindle: note the arms and bridges projecting from most MTs. Bar = 0.2 um. (From McDonald et al., 1979).

in Surirella calcarata with such precision that subsequent electron microscope observations served more to confirm his results than supplant them. Pickett-Heaps et al. (1984) have provided the modern reader an opportunity to appreciate these detailed observations by translating and reprinting some of the text and figures from the 1896 monograph.

Ultrastructural studies of algal (and diatom) mitosis began appropriately enough with the work of Irene Manton, one of the pioneers of modern cell biology (see Manton, 1978). Manton and her co-workers (Manton et al., 1969a,b, 1970a,b) described spindle structure in the diatom Lithodesmium in greater detail than any previous study of mitotic ultrastructure. By counting spindle microtubules in selected serial sections, they provided evidence which strongly suggested that the spindle was composed of two half spindles which interdigitated in a zone of overlap.

In the mid-1970's Pickett-Heaps' laboratory resumed the study of mitotic spindle ultrastructure in diatoms. It was clear from the papers on Diatoma (Pickett-Heaps et al., 1975) and Melosira (Tippit et al., 1975) that the diatoms were excellent material for studying the mechanism of anaphase spindle elongation. Movement of chromosomes to the poles (anaphase A) was spatially and temporally separate from the separation of the poles themselves. Furthermore, the central spindle had that kind of regularity (Figs. 1-2) which is useful in analyzing cell motility phenomena. Longitudinal sections suggested that the central spindle was composed of two overlapping half spindles which moved apart during spindle elongation. By mapping the distribution of spindle MTs in serial cross-sections (Fig. 3), McDonald et al. (1977) were able to show that this was indeed the case. Near neighbor analysis of the interdigitating half spindle MTs (McDonald et al., 1979; McIntosh et al., 1979) showed that antiparallel MTs (those originating from opposite poles) were preferentially associated in the zone of overlap at a spacing of 42 nm (center to center). The fact that some MTs were in a square-packed arrangement further suggests that some bridge molecule is holding the MTs in this configuration. Without this kind of energetic input into the system one would expect to see hexagonally-packed MTs. Although there are clearly bridges between MTs (Fig. 2), the inconsistency of their presence suggests that they are probably labile and easily destroyed during processing for electron microscopy.

Leslie and Pickett-Heaps (1983,1984) have used UV microbeam irradiation of diatom central spindles to investigate the mechanism of spindle elongation. When the spindle was severed near the pole, the remnant which included the overlap would continue to elongate if the cell was in anaphase. Interestingly, metaphase cells treated the same way would not elongate. The authors suggest that

spindle elongation is a discrete process which needs to be
"activated" in anaphase. A clear implication of their
results is that the motor for spindle elongation is located
in the zone of overlap. These studies also show that half
spindle MTs depolymerize from the overlap toward the poles,
confirming the observations of Sorrano and Pickett-Heaps
(1982) on the post-mitotic disassembly of spindles in
Pinnularia.

THE STEPHANOPYXIS TURRIS SYSTEM

Development of the functional in vitro model.

Given their high degree of order, we reasoned that
diatom spindles would be excellent for developing a
functional in vitro model system. As in axonemes, the cross-
linked bundles of MTs would probably stand more vigorous
isolation and purification procedures and yet remain
capable of motility when an energy source such as ATP was
added to the system. After screening numerous species from
several collections, we decided on Stephanopyxis turris
because of its natural mitotic synchrony in culture, the
fact that the spindle was large enough for critical light
microscope observations, yet not so large to prohibit
quantitative structural analyses at the EM level.
Stephanopyxis turris is a marine centric diatom which
grows well in culture in Guillard's modified F/2 medium
(Guillard, 1975). Interphase nuclei are located at the end
walls and prior to mitosis they migrate along the side walls
to a central position up against the girdle bands (Fig. 4).
This migration seems to be mediated by cytoplasmic
microtubules (Wordeman et al., 1986). Spindle formation is
comparable to other diatoms and metaphase spindle structure
is also typical, with an overlap zone of about 20-25% the
total spindle length (McDonald et al., 1986). Spindle
elongation in vivo takes place at a rate of approximately
1.5 um/min. and the extent of elongation is 2-3 times the
length of the overlap zone. The number of half spindle MTs
in diatoms varies widely, from twenty or less in Fragilaria
(Tippit et al., 1978) to five hundred or more in Pinnularia
(Pickett-Heaps et al., 1978). S. turris spindles number
about 300 per half spindle.

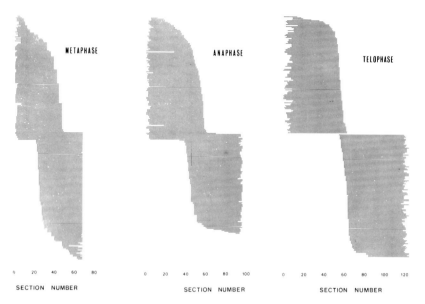

SECTION NUMBER   SECTION NUMBER   SECTION NUMBER

Figure 3. The distribution of individual MTs from central spindles at metaphase, anaphase and telophase: each MT tracked through serial sections, then arranged along the spindle axis (x-axis) according to point of origin and termination: y-axis has no significance. The relative displacement of the half spindles between metaphase and telophase suggests a sliding apart. (From McDonald et al., 1977).

Cells growing at a density of about 10^{-6} cells/l can have a natural mitotic peak at about 2 h after the onset of the dark phase in a 18:6 h light-dark cycle. The number of cells entering mitosis at this time can be increased by incubating the cells in 5 x 10^{-7} M nocodazole at the beginning of the dark cycle, then washing it out after 3h. Cells prepared in this way in multiple 9 liter carboys are the starting material for the experiments described in the text which follows. The most up to date versions of culture and synchrony conditions are described in the papers by Masuda and Cande (1987, Masuda et al., 1988).

About 20 min after removal of the nocodazole, 20-30% of the cells contain mitotic spindles in metaphase or early anaphase. Cells are concentrated by filtration and centrifugation then fractionated by several passes in a

Dounce homogenizer. Because diatoms have a delicate silicon
wall as opposed to a fibrous cellulosic wall, the
fractionation can be relatively gentle and the nuclei with
their associated spindles easily separated from most of the
other cellular components by filtration and centrifugation.
The main contaminants in an isolated spindle preparation are
interphase nuclei, chloroplasts and fragments of cell wall.
To stabilize the mitotic spindle microtubules, glycerol and
taxol can be added to the homogenization medium as well as
proteolysis and phosphatase inhibitors which help maintain
the integrity of other spindle-associated proteins. Spindles
remain embedded in a mass of chromatin unless the chromatin
is removed by DNAase treatment. For most experiments it is
useful to leave the chromatin attached because it acts as a
"glue" when spindles/nuclei are spun down onto coverslips.
A coverslip containing dozens to hundreds of attached
spindles is the starting material for most of the
reactivation and immunolocalization studies described here.

Physiological studies.

     When isolated spindles are incubated in reactivation
medium containing ATP (Cande & McDonald, 1985) they will
elongate to the extent of the overlap zone length, i.e., ca.
20-25% original spindle length (Fig. 5). In polarization
optics, the overlap zone goes from a highly birefringent
area to one with little or no birefringence. The spindles
appear to have a "gap" in the middle. Measurements of
populations of reactivated spindles shows a correlation
between increase in spindle length and the presence of a gap
(Cande & McDonald, 1986). Thus, by assaying for the presence
or absence of gaps, we can measure with relative ease the
effects of different physiological treatments on large
numbers of spindles. This assay increases our sample size
considerably and therefore gives us more confidence in the
accuracy and reproducibility of our results.
     To investigate the physiological and biochemical
nature of the "motor" which drives spindle elongation we
treated isolated spindles with reagents which inhibit the
known mechanochemical ATPases dynein, myosin and kinesin
(Cande et al., 1988). Although the spindles have several
properties which are dynein-like and one that is like
kinesin, we believe that the motor is probably biochemically
distinct from any of the known motility ATPases.

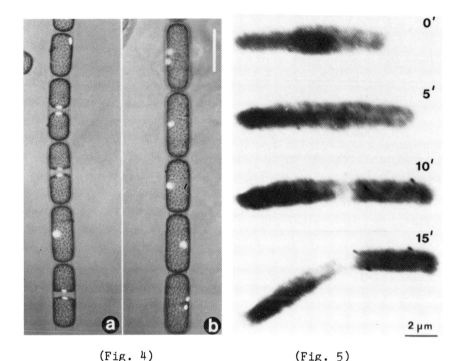

(Fig. 4)   (Fig. 5)

Figure 4a. Strands of DAPI-stained S. turris cells in
different stages of nuclear migration and mitosis: bright
spots are nuclei. Cell at top is in nuclear migration:
second cell from bottom has centered mitotic nucleus; other
cells are newly cleaved cell pairs. Figure 4b. A strand of
mitotic cells. Bar = 100 um. (From Wordeman et al., 1986).

Figure 5. Isolated spindle viewed with polarization optics
with negative compensation, just before ATP addition (0 min)
and after addtion of 1 mM ATP (5, 10, 15 min): note decrease
in birefringence in zone of MT overlap during spindle
elongation. Bar = 2 um. (From Cande & McDonald, 1985).

Biochemical and immunological studies of phosphorylation.

        Isolated spindle preparations include a number of
enzymes which will affect reactivation if they are not
controlled. For example, proteolysis is a real problem and a
cocktail of proteolytic inhibitors (Cande et al., 1983) is
necessary to retain spindle reactivation capabilities.
Phosphatases and kinases are probably present also. Without

phosphatase inhibitors, the number of spindles which
reactivate is greatly reduced. Wordeman and Cande (1987)
took advantage of this fact to study the role of
phosphorylation in spindle reactivation as well as to
increase the total number of spindles which would elongate
in the presence of ATP. Spindles isolated under conditions
which were permissive for endogenous phosphatase activity
but not spindle elongation could be "rescued" by incubating
them in low levels of ATPgammaS. Presumably, endogenous
kinases in the spindle preparations used the non-
hydrolyzable analog to irreversibly phosphorylate spindle
proteins which are necessary for achieving spindle
elongation in vitro. Including beta-glycerophosphate as a
competitive inhibitor of endogenous phosphatases will also
improve reactivation percentages.

A monoclonal antibody to the thiophosphate group
(Gerhart et al., 1985) was used to confirm that proteins
were being thiophosphorylated and to localize them in the
spindle (Wordeman and Cande, 1987). When spindles are
incubated in low concentrations of ATPgammaS, antibody
labelling is mainly in the overlap zone (Wordeman & Cande,
1987) with some staining of poles and presumptive

EXOGENOUS TUBULIN   NATIVE TUBULIN

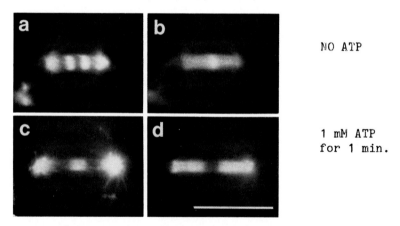

NO ATP

1 mM ATP
for 1 min.

Figure 6. Exogenous biotinylated neurotubulin adds to the
pole-distal ends of each half spindle (a) and moves together
into a "new" overlap (c) as the result of ATP-induced
sliding. Double staining with fluorescent antibodies against
biotin (a,c) and plant tubulin (b,d). Bar = 10 um. (From
Masuda et al., 1988)

kinetochores. Immunoblots of spindle protein extracts show that the anti-thiophosphate antibody reacts primarily with a protein of approximately 205 kD (Wordeman & Cande, 1987). Because the 205 kD protein is released into the supernatent when spindle microtubules are depolymerized, it may be a microtubule-associated protein (MAP). Alternatively, or in addition, this protein may be part of the regulatory machinery or a component of the mechanochemical enzyme complex which generates the force for spindle elongation. Regardless of its exact nature, we know that it must be phosphorylated in order for spindle elongation to take place in vitro (Wordeman & Cande, 1987). An antibody to mitosis-specific phosphorylated proteins in HeLa cells (Davis et al., 1983) also reacts with the 205 kD protein as well as a number of mitotic spindle structures in a wide variety of cell types (Vandre et al., 1986; Vandre & Borisy, 1986). Clearly, the phosphorylation state of mitotic proteins is important to the regulation of chromosome and spindle movements. It is also clear that diatom spindles are excellent material with which to study this aspect of mitosis.

Microtubule polymerization and spindle elongation

In vitro, the spindle of S. turris elongates to the extent of the overlap zone length. However, in vivo, it may elongate many times the length of the original overlap zone (McDonald et al., 1986). This means that the average half spindle length must increase due to MT polymerization, or that the half spindle MTs must "telescope" apart. Masuda and Cande (1987; Masuda et al., 1988) have studied this problem with the S. turris model system. Using biotinylated tubulin as a marker for new tubulin addition, they found that incorporation takes place in the region of the overlap (Fig. 6). This is contrary to the prediction of Pickett-Heaps et al. (1986) that incorporation takes place at the minus, or polar ends. Although there can be addition of tubulin to the poles in S. turris, it is non-specific and can be eliminated if taxol is removed from the polymerization medium (Masuda and Cande, 1987). Incorporation of neurotubulin is not ATP-dependent and can therefore be uncoupled from spindle elongation (Masuda et al., 1988) and lead to unusually large overlap zones prior to pole separation. Upon addition of ATP, the spindle elongates a distance equal to the length of the new overlap. Toward the end of elongation, this means

that the overlap is composed entirely of neurotubulin.
Electron microscopy shows that the new tubulin adds on to
the ends of existing microtubules (Masuda et al., 1988).
        Double immunolabelling with anti-biotin and anti-
thiophosphate antibodies of spindles pre-incubated in
neurotubulin shows that the thiophosphorylated proteins
remain in the zone of overlap even if it is made up of only
the new, biotinylated MTs (Cande et al., 1988). In the
absence of new MTs, the distribution of thiophosphorylated
proteins would decrease as the overlap decreased. Its
behavior parallels the behavior of the osmiophilic "fuzz"
seen in the overlap zone by electron microscopy and may be a
component of the fuzz. The fact that it remains in place
even when new MTs comprise the overlap suggests that it is
associated with an overlap matrix material rather than the
microtubules themselves.

Models of spindle elongation

        Our results have eliminated for diatoms several
possible mechanisms for the separation of spindle poles or
spindle elongation. The forces cannot be due to pulling on
the poles as some have suggested for fungi (Aist & Burns,
1981) because the isolated spindles have no cytoplasm beyond
the poles (Cande & McDonald, 1985). Elongation cannot be due
to MT polymerization because spindles will elongate without
any tubulin present. When tubulin is present, it adds to the
overlap region but spindles do not elongate. Because the
spindles are normally embedded in chromatin, it might be
thought that swelling of the chromatin when ATP is added
could cause the half spindles to separate. However, the
chromatin can be digested away with DNAase and the spindles
will still reactivate (McDonald et al., 1986).
        Everything that we have discovered about the
structure, biochemistry and physiology of the S. turris
model suggests that the motor for spindle elongation is
located in the zone of overlap. Furthermore, its proper
functioning requires that a 205 kD protein be phosphorylated
(Wordeman & Cande, 1987). The distribution of this protein
in spindles which have been reactivated in the presence of
exogenous tubulin also suggests that the motor is not
permanently attached to the microtubules but may be part of
the matrix between microtubules. The fact that exogenous
tubulin can be moved through the overlap zone is consistent
with this idea. Alternatively, one can imagine that the

motor could somehow move along the microtubules as they slid out of the midzone, however, this idea is less appealing than the matrix model. Finally, the results of inhibitor studies suggest that the motor is probably not one of the standard mechanochemical ATPases - dynein, myosin or kinesin. Although there are some dynein-like parallels in the pharmacology, the polarity of the sliding MTs is wrong. In cilia and flagella the adjacent MTs are parallel while in the diatom they are antiparallel.

PROSPECTUS

The diatom central spindle is already the most well-characterized mitotic spindle structurally and with the development of a functional in vitro system it has great promise for studying the physiology and biochemistry of anaphase spindle elongation. Although this is only one of the several different ways in which chromosomes move during mitosis and not as striking as the movement of chromosomes to the poles, the chances of identifying the "motor" for spindle elongation seem better than for any other type of mitotic movement. We know where to look - the overlap region - and we know that a phosphorylated 205 kD protein is probably involved. Preliminary results with a fast-growing diatom (Cylindrotheca) are encouraging because we can get decent yields of spindle extract for biochemistry and possible antibody production. If further work with the Cylindrotheca system is successful, the next few years of mitosis research using algal systems will be very productive and exciting.

REFERENCES

Aist JR, Berns MW (1981). Mechanisms of chromosome separation during mitosis in Fusarium (Fungi Imperfecti): new evidence from ultrastructural and laser microbeam experiments. J Cell Biol 91:446-458.
Bajer A, Mole-Bajer J (1972). Spindle dynamics and chromosome movements. Int Rev Cytol Suppl 3:1-271.
Cande WZ (1986).Reactivation of mitosis in vitro. Trends in Biochem Sci 11:447-449.
Cande WZ, Tooth PJ, Kendrick-Jones J (1983). Regulation of contraction and thick-filament assembly-disassembly in glycerinated verterbrate smooth muscle cells. J Cell Biol 97:1062-1071.

Cande WZ, McDonald KL (1985). In vitro reactivation of anaphase spindle elongation using isolated diatom spindles.
Nature (Lond) 316:168-170.

Cande WZ, McDonald KL (1986). Physiological and ultrastructural analysis of elongating mitotic spindles reactivated in vitro. J Cell Biol 103:593-604.

Cande WZ, Baskin T, McDonald KL, Masuda H, Wordeman L
(1988). Anaphase spindle elongation studied by in vitro
methods. In Warner FD, McIntosh JR (eds): "Force Production and Microtubule-coupled Movement," in press.

Cassimeris LU, Walker RA, Pryer NK, Salmon ED (1987)
Dynamic instability of microtubules. Bioessays 7:149-154.

Davis FM, Tsao TY, Fowler SK, Rao PN (1983). Monoclonal
antibodies to mitotic cells. Proc Natl Acad Sci USA 80:
2926-2930.

Euteneuer U, McIntosh JR (1980). Polarity of midbody and
phragmoplast microtubules. J Cell Biol 87:509-515.

Euteneuer U, McIntosh JR (1981). Structural polarity of
kinetochore microtubules in PtK1 cells. J Cell Biol 89:
338-345.

Gerhart J, Cyert M, Kirschner M (1985). M-phase promoting
factors from eggs of Xenopus. Cytobios 43:335-347.

Guillard RRL (1975). Culture of phytoplankton for feeding
marine invertebrates. In Smith WL, Chanley MH (eds):
"Culture of Marine Invertebrate Animals," New York:
Plenum, pp 29-60.

Heath IB (1980). Variant mitoses in lower eukaryotes: indicators of the evolution of mitosis? Int Rev Cytol 64:
1-80.

Heidemann SR, McIntosh JR (1980). Visualization of the
structural polarity of microtubules. Nature 286:517-519.

Inoue S, Stephens RE (eds)(1975). "Molecules and Cell
Movement," New York: Raven Press, pp 1-450.

Lauterborn R (1896). Untersuchungen uber Bau, Kernteilung
und Bewegung der Diatomeen. Leipzig: W. Engelmann,
pp 1-165.

Leslie RJ, Pickett-Heaps JD (1983). Ultraviolet microbeam
irradiation of mitotic diatoms. Investigation of spindle
elongation. J Cell Biol 96:548-561.

Leslie RJ, Pickett-Heaps JD (1984). Spindle dynamics following ultraviolet microbeam irradiations of mitotic
diatoms. Cell 36:717-727.

Luykx P (1970). Cellular mechanisms of chromosome distribution. Int Rev Cytol Suppl 2:1-173.

Manton, I (1978). Recollections in the history of electron

microscopy. Proc Roy microscop Soc 13:45-57.
Manton, I, Kowallik K, von Stosch HA (1969a). Observations
of the fine structure and development of spindle at
mitosis and meiosis in a marine diatom (Lithodesmium
undulatum). I. Preliminary survey of mitosis in spermio-
genesis. J Microsc (Oxford) 89:295-320.
Manton I, Kowallik K, von Stosch HA (1969b). II. The early
meiotic stages in male gametogenesis. J Cell Sci 5:271-298
Manton I,Kowallik K, von Stosch HA (1970a). III. The later
stages of meiosis I in male gametogenesis. J Cell Sci
6:131-157.
Manton I, Kowallik K, von Stosch HA (1970b). IV. The second
meiotic division and conclusion. J Cell Sci 7:407-444.
Masuda H, Cande WZ (1987). The role of tubulin polymerization during spindle elongation in vitro. Cell 49:193-202.
Masuda H, McDonald KL, Cande WZ (1988). The mechanism of
anaphase spindle elongation: uncoupling of tubulin incorporation and microtubule sliding during in vitro
spindle reactivation. J Cell Biol 107:623-633.
McDonald KL, Pickett-Heaps JD, McIntosh JR, Tippit DH
(1977). On the mechanism of anaphase spindle elongation
in Diatoma vulgare. J Cell Biol 74:377-388.
McDonald KL, Edwards MK, McIntosh JR (1979). Cross-sectional
structure of the central mitotic spindle of Diatoma
vulgare. Evidence for specific interactions between anti-
parallel microtubules. J Cell Biol 83:443-461.
McDonald KL, Euteneuer U (1983). Studies on the structure
and organization of microtubule bundles in the interzone
of anaphase and telophase PtK1 cells. J Cell Biol 97:88a.
McDonald KL, Pfister K, Masuda H, Wordeman L, Staiger C,
Cande WZ (1986). Comparison of spindle elongation in vivo
and in vitro in Stephanopyxis turris. J Cell Sci Suppl
5:205-227.
McIntosh JR, McDonald KL, Edwards MK, Ross BM (1979). Three-
dimensional structure of the central mitotic spindle of
Diatoma vulgare. J Cell Biol 83:428-442.
Nicklas RB (1971). Mitosis. Adv Cell Biol 2:225-294.
Pickett-Heaps JD (1974)." Green Algae," Sunderland, MA:
Sinauer, pp 1-606.
Pickett-Heaps JD, McDonald KL, Tippit DH (1975). Cell divi-
sion in the pennate diatom Diatoma vulgare. Protoplasma
86:205-242.
Pickett-Heaps JD, Tippit DH (1978). The diatom spindle in
perspective Cell 14:445-467.
Pickett-Heaps JD, Tippit DH, Andreozzi JA (1978). Cell
division in the pennate diatom Pinnularia. II. Later

stages in mitosis. Biol Cell 33:79-84.

Pickett-Heaps JD, Schmid A-MM, Tippit DH (1984). Cell division in diatoms. A translation of part of Robert Lauterborn's treatise of 1896 with some modern confirmatory observations. Protoplasma 120:132-154.

Pickett-Heaps JD, Tippit DH, Cohn SA, Spurck TP (1986). Microtubule dynamics in the spindle. Theoretical aspects of assembly/disassembly reactions in vivo. J theoret Biol 118:153-169.

Schrader F (1953). "Mitosis" 2nd ed, New York: Columbia University Press, xii + 170 pp.

Soranno T, Pickett-Heaps JD (1982). Directionally controlled spindle disassembly after mitosis in the diatom Pinnularia. Eur J Cell Biol 26:234-243.

Telzer BR, Haimo LT (1981). Decoration of spindle microtubules with dynein: evidence for uniform polarity. J Cell Biol 89:373-378.

Tippit DH, McDonald KL, Pickett-Heaps JD (1975). Cell division in the centric diatom Melosira varians. Cytobiologie 12:52-73.

Tippit DH, Schulz D, Pickett-Heaps JD (1978). Analysis of the distribution of spindle microtubules in the diatom Fragilaria. J Cell Biol 79:737-763.

Vandre DD, Borisy GG (1986). The interphase-mitosis transformation of the microtubule network in mammalian cells. In Ishikawa H, Hotano S, Sato H (eds): "Cell Motility: Mechanisms and Regulation," New York: Alan R. Liss, pp 389-401.

Vandre DD, Davis FM, Rao PN, Borisy GG (1986). Distribution of cytoskeletal proteins sharing a conserved phosphorylated epitope. Eur J Cell Biol 41:72-81.

Wordeman L, McDonald KL, Cande WZ (1986). The distribution of cytoplasmic microtubules throughout the cell cycle of the centric diatom Stephanopyxis turris: their role in nuclear migration and positioning the mitotic spindle during cytokinesis. J Cell Biol 102:1688-1698.

Wordeman L, Cande WZ (1987). Reactivation of spindle elongation in vitro is correlated with phosphorylation of a 205 kD spindle-associated protein. Cell 50:535-543.

ACKNOWLEDGMENTS

We wish the thank Drs. Linda Wordeman and Hirohisa Masuda for contributing illustrations for this paper. Manuscript preparation was supported by NIH grants 1 R24 RR03744 to KM and GM23238 to WZC.

Algae as Experimental Systems pages 19–37
© 1989 Alan R. Liss, Inc.

ALGAL CENTRIN: CALCIUM-SENSITIVE CONTRACTILE ORGANELLES

Jeffrey L. Salisbury

Center for NeuroSciences, Case Western Reserve University, School of Medicine, Cleveland, Ohio 44106

"We must therefore ascribe to living cells, beyond the molecular structure of the organic compounds that they contain, still another structure of different type of complication; and it is this which we call by the name of organization."
Brücke, 1861

The origin and maintenance of cell form (cellular morphogenesis) in eucaryotes is a complex derivative of interactions which take place during development between the cellular environment and the principal cytoskeletal systems of the cytoplasm: actin-based microfilaments, intermediate filaments, and the microtubule complex. In animal cells, microfilament-based cortical cytoskeletal structures are, in large part, responsible for surface related motility, for the control of membrane topology and the distribution of certain surface proteins, and for overall cell shape. In addition, the actin-based cytoskeleton is responsible for general properties of cell contractility, and during division, for cell cleavage. While certain of these properties have been eliminated altogether or supplanted by cell wall functions in algal cells, others clearly remain dependent on an actin containing cell cortex; for example cytoplasmic streaming in algal filaments (LaClaire, 1988), mating structure activation in *Chlamydomonas* (Detmers *et al.*, 1985), and cell polarization and cleavage in fucoid zygotes (Brawley & Robinson, 1985; Quatrano & Kropf, 1988). Intermediate filaments represent a family of related proteins which generally confer properties of resilience and elasticity to the cytoplasm of animal cells. Intermediate filaments also appear to serve a structural (space-filling) role in certain highly differentiated animal cell types; examples include the keratin filaments of epithelial cells, and neurofilaments of neuronal processes such as axons. The presence and function of intermediate filaments in algal cells is less clear, although simple eucaryotes and plant cells have been reported to contain proteins antigenically related to intermediate filaments (Dawson *et al.*, 1985; Miller *et al.*, 1985; Numata *et al.*, 1985). The microtubule complex is a ubiquitous element of cytoplasmic organization in eucaryotic cells which defines a "coordinate geometry" for the cytoplasm. Microtubules participate in directed intracellular motility such as chromosome movement and certain vesicular transport processes, as well as in ciliary motility. Present understanding of cytoplasmic structure views these cytoskeletal filament systems in a collective sense: as a cytomatrix (Porter, 1984). In addition, the cytomatrix comprises a host of associated proteins which serve to bind, crosslink, cap, sever, buffer, organize, and move elements of the cytoskeletal framework. Thus, the cytomatrix can be thought of as a supraorganelle of cytoplasmic structure, which is both exquisitely ordered and highly dynamic.

Insight into the molecular basis for the organization of the cytoplasm in eucaryotic cells comes from recent technical advances in electron microscopy, cell fractionation, the use of genetic and immunological tools, and protein biochemistry. Cell biologists have recognized the gross morphological and functional correlates of the principal cytoskeletal domains for over one hundred years (Wilson, 1925). Indeed, many of the questions which concerned cell biologists of the past century remain of interest today, and not the least of these querries is the nature and origin of structural and functional polarity of cells and cell motility.

Van Beneden (1883) first formally defined the structural polarity observed in many cell types as a "cell axis" which is indicated by the position of the centrosome, containing the centrioles, with reference to the cell nucleus: the cell axis passes through the center of both (Fig. 1). The centrioles and a variable portion of differentiated cytoplasm surrounding them, together known as the centrosome, were viewed as the "dynamic center of the cell" by early cell biologists. The relationship between the centrosome and the cell axis is particularly evident in highly motile cells such as macrophages as well as in epithelia. Certain monads, including unicellular algal cells, illustrate extreme examples of the cell axis in that their cellular organization is almost bilaterally symmetrical about the plane of the flagellar apparatus and nucleus (Fig. 1). *Chlamydomonas* and other green algal unicells are particularly good examples of this type of structural organization. The algal flagellar basal apparatus and mammalian centrosome can be viewed as structural and functional equivalents in that both organelle complexes represent the primary microtubule organizing center of interphase cells and during mitosis both occur at the mitotic spindle poles and harbor centrioles or basal bodies. Likewise, numerous motility phenomena are centered on these organelle complexes in both animal and algal cells, including; flagellar motility, flagellar root contraction, and centrosome position and equipartition into daughter cells during mitosis. Therefore algal cells represent model systems for the structural, biochemical, and genetic characterization of the eucaryotic centrosome.

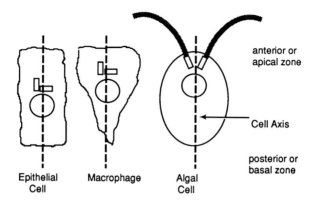

Figure 1. The cell axis defines cell polarity and is indicated by a line drawn through the centrosome, containing the centrioles, and the nucleus. Algal cells such as *Chlamydomonas* illustrate an extreme example of the cell axis.

In this review I will address findings from our laboratory and from other workers concerning the centrosome-associated protein centrin. Centrin is a 20,000 MW, acidic, phosphoprotein, that was first identified in algal cells (Salisbury *et al.*, 1984) as a component of contractile flagellar roots (Salisbury & Floyd, 1978). Recent studies utilizing poly- and monoclonal antibodies directed against centrin epitopes have demonstrated the ubiquitous occurrence of centrin homologues in all eucaryotic cells studied to date, including algal, protozoan, higher plant, and mammalian cell types (Salisbury *et al.*, 1986; Schulze, *et al.*, 1987; Baron & Salisbury, 1988). In all, centrin is associated with the basal bodies of the flagella, centrioles, centrosome, or mitotic spindle poles of the cell. In *Chlamydomonas* and many other algal unicells, centrin links the basal bodies or centrioles to the nucleus (Wright *et al.*, 1985; Salisbury *et al.*, 1987; Schulze *et al.*, 1987). Contraction of centrin containing fibers results in reorientation of the flagellar basal bodies (McFadden, *et al.*, 1987), flagellar excision (Sanders & Salisbury unpublished observations), nuclear movement and nuclear shape changes (Salisbury *et al.*, 1987), and in dynamic behavior of the mitotic spindle poles and spindle matrix, particularly during preprophase and at the metaphase/anaphase transition (Salisbury *et al.*, 1988; Baron & Salisbury, 1988). Centrin dynamics during the cell cycle correlate with the "flexible centrosome" as defined by Mazia (1984), and therefore studies of centrin biochemistry, physiology, and behavior in algal model systems are contributing to a broader understanding of this enigmatic organelle complex which has fascinated cell biologists for so many years (Wilson, 1925; cf., Wheatley, 1982).

THE ALGAL FLAGELLAR APPARATUS

Flagella are motile organelles that propel cells through water (Satir & Ojakian, 1979) and they act as specialized recognition and adhesion organelles during sexual mating (Goodenough & Adair, 1988; vanden Ende, 1988). In addition, they increase the surface area of the cell membrane which is exposed to the environment (e.g., the combined surface area of the two flagella of *Chlamydomonas* is about 10% of the total cell surface; Lewin *et al.*, 1982); thus, flagella serve as the walled algal cell's window on the world. Motile algal cells are capable of a large number of behavioral responses (i.e., phototaxis and photophobic response, sexual mating, gliding, flagellar excision and regrowth). Since algae do not reason, the execution of appropriate action by an individual cell must, in part, be preprogrammed into a finite set of behavioral subroutines that are determined by structural information embodied in the organization of the flagella and the flagellar basal apparatus.

The early ultrastructural work of Manton (Manton, 1956) on the structure of plant flagella and associated organelles presaged the focus of much of the current work in this area. Manton and her coworkers established that the substructure of the flagellar axoneme consists of nine outer doublet fibers (microtubules) arranged in a cylinder surrounding two central fibers (also microtubules); and that this basic structure, surrounded by an extension of the cell membrane, is ubiquitous among flagellated and ciliated eucaryotes (Manton, 1956; Fawcett & Porter, 1954). Our understanding of the details of the structure, composition, genetics and motility of flagella has progressed rapidly since these early studies (Huang, 1986; Lefebvre & Rosenbaum, 1986; Ringo, 1967; Witman *et al.*, 1972). There have been a number of excellent recent reviews on the structure of the flagellar apparatus in green algal cells (Melkonian, 1980, 1982, 1984; Pickett-Heaps, 1975; and cf., Cox, 1980) to which I refer the reader for details. Here, I will reiterate only general features and those points which relate to our understanding of centrin-based structures associated with the flagellar apparatus.

Flagella are complex microtubule-based organelles composed of several hundred distinct polypeptides. Each outer doublet microtubule of the axoneme is composed of a complete microtubule (A subfiber, with 13 protofilaments) and a partial microtubule (B subfiber, with 11 protofilaments) which shares a portion of the wall of the A microtubule. Dynein arms are mechano-chemical transducers that extend from each A subfiber toward the adjacent B subfiber. Dynein arms are responsible for active sliding of adjacent doublets in a base-to-tip direction during flagellar motility (Sale & Satir, 1977). The sliding motion is converted to an axonemal bend by the action of accessory structures that include radial spokes and nexin links. Flagellar motility is dependent on ATP and is regulated in a complex fashion; active sliding along one half of the axonemal cylinder is switched on during the power or effective stroke while the opposite half of the axonemal cylinder is active during the recovery or return stroke (cf., switch point hypothesis; Satir, 1982). In addition, when two flagella are present, each may operate under differential calcium sensitive control mechanisms which effect both the rate and form of flagellar motion (Kamiya & Witman, 1984; Omoto & Brokaw, 1985).

Flagellar Basal Apparatus: Basal Bodies

The flagellar axoneme is anchored in the cell at the basal body where the microtubules of the axoneme are continuous with microtubules of the basal body. Basal bodies contain additional partial microtubules (C subfibers) along the outer wall of each B subfiber; thus each basal body is a cylinder (typically on the order of 0.5 μm long) of nine triplet microtubules (Ringo, 1967). In *Chlamydomonas* the pair of basal bodies lie in nearly the same plane at approximately 80-90° relative to one another and with a 180° of rotational symmetry (Ringo, 1967; Hoops & Witman, 1983). The basal bodies are connected to one another by a striated distal fiber located midway along the length of their inner walls (Fig. 2c,e-f), and by a pair of fibers at their proximal ends. The distal connecting fiber is complex and bilaterally symmetrical with a pair of dark striations near the center, a fine striation and another pair of dark striations on either side (Ringo, 1967).

In many organisms the transition zone between the axoneme and basal body is specialized for flagellar excision or autotomy ("self-cutting"; Blum, 1971; Lewin *et al.*, 1982). Flagellar excision occurs in response to adverse environmental perturbations such as temperature extremes, pH shock or alcohol treatment, and mechanical shear. It has been suggested that flagellar excision has survival value for the organism; by casting off their flagella cells expose less surface to the environment thereby reducing the area of membrane susceptible to chemical or physical trauma (Blum, 1971; Lewin *et al.*, 1982). Flagellar excision is an active (Lewin *et al.*, 1982) process that is under genetic control; *Chlamydomonas* mutants defective in the excision process have been isolated (Lewin & Burrascano, 1983). Flagellar excision is calcium-dependent and involves structural changes in the distal-most region of the transition zone (Huber *et al.*, 1986; Lewin & Lee, 1985). The transition zone is characterized by an H-shaped central cylinder which, in cross-section, appears as a stellate fibrous structure that links the central cylinder to the A microtubules of the axonemal doublets (Lang, 1963; Ringo, 1967). At the time of flagellar excision, the stellate structure contracts resulting in the application of a torque load and/or shear force on the axonemal doublets thereby severing the microtubules of the axoneme from the transition zone (Sanders &

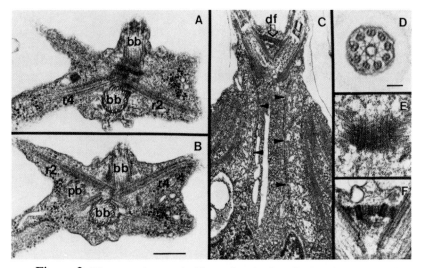

Figure 2. Electron micrographs illustrating the flagellar basal apparatus of *Chlamydomonas*. A, B) Consecutive serial transverse sections through the apical region of a cell revealing cruciate microtubular rootlets (*r 2, r 4*), two basal bodies (*bb*), and a probasal body (*pb*): bar = 0.25 μm. C) Section through the plane of the basal apparatus revealing the distal fiber (*df*), basal bodies, the H-piece of a transition zone, and centrin-based roots (arrow heads) leading to the nucleus. D) Cross section through a transition zone showing a stellate structure: bar = 100nm. Distal fibers sectioned *en face* (E) and longitudinally (F). [C, Reprinted with permission from Salisbury *et al.*, (1988) J Cell Biol 107:635.]

Salisbury, unpublished observations). We are particularly interested in the excision process because it is calcium sensitive and because immuno-gold localization studies show that the stellate structure of the H-piece contains the calcium-sensitive contractile protein centrin (Sanders & Salisbury, unpublished observations). The mechanism of flagellar excision, therefore, appears to be based on a calcium-mediated contraction of centrin-containing fibers within the transition zone. Following flagellar excision, cells may grow new flagella utilizing both a pool of pre-existing precursors and newly synthesized protein (cf. Lefebvre & Rosenbaum, 1986). This will be discussed in greater detail below.

Basal Body Development

The two basal bodies of *Chlamydomonas* are developmentally distinct. Basal body duplication in the algae is semiconservative (Aitchison & Brown, 1986; Melkonian *et al.*, 1987); one basal body each of the pre-existing pair and one newly developed basal body is passed on to each daughter cell during cell division. Development of probasal bodies in *Chlamydomonas* starts at mid-metaphase and is finished by early telophase (Cavalier-Smith, 1974; Gaffal, 1988) . The probasal bodies persist without elongation throughout cytokinesis and the subsequent interphase period until the onset of the next mitosis (Cavalier-Smith, 1974; Gaffal, 1988; Gould, 1975). During prophase, the basal body/microtubular root complex (see below) becomes partitioned in a plane perpendicular to the distal connecting fiber; therefore each of the separated halves of the original basal apparatus consists of one older and one younger basal body (Gaffal, 1988). Dutcher and coworkers (Huang *et al.*, 1982; Dutcher *et al.*, 1988) have isolated a series of *Chlamydomonas* mutants (*uni*) in which the two basal bodies show uniquely exaggerated maturation properties; the basal body cis to the eyespot fails to grow a flagellum, while the basal body trans to the eyespot grows a flagellum. In addition, the transition zone of the trans basal body has an altered H-piece. Following cell division in these mutants, each daughter cell has the ability to grow only a single flagellum (again, trans to the eyespot). The studies of Melkonian and coworkers (Melkonian *et al.*, 1987) on maturation and conversion of long and short flagella in *Nephroselmis* suggest that flagella that are newly formed at the beginning of a cell cycle change to the other flagellar type during the following cell cycle. Taken together these observations suggest that the cis basal body becomes refractory to the *uni* mutant effector during the mitotic cycle thereby gaining the ability to grow a flagellum during the ensuing cell cycle and that the cis basal body is converted to a trans basal body near the end of each cell cycle.

Distal Fiber Contraction

McFadden and coworkers (1987) have shown that the distal fiber that connects the two adjacent basal bodies to one another is contractile in the green alga *Spermatozopsis*. The distal fiber is composed, at least in part, of the calcium-modulated protein centrin (McFadden *et al.*, 1987). Calcium sensitive contraction of the distal fiber is responsible for flagellar reorientation during the photophobic response in *Spermatozopsis* (McFadden *et al.*, 1987). The distal fiber of *Chlamydomonas* is also centrin-based (Salisbury *et al.*, 1987, 1988). Elevated free calcium levels have been shown to result in flagellar reorientation of the isolated flagellar apparatus of *Chlamydomonas* (Hyams & Borisy, 1975, 1978). However, there are no observations of flagellar reorientation in living *Chlamydomonas* cells. This may be due to the presence of a rigid cell wall and flagellar tunnels, which preclude gross physical displacement of the basal bodies. Perhaps in *Chlamydomonas* a more subtle mechanical stimulation of the basal bodies results in the signaling of flagellar responses and in setting the coordination of the two flagellar beat cycles (Salisbury & Floyd, 1978; Salisbury *et al.*, 1981).

Microtubule Rootlets

A microtubule-based cortical cytoskeleton converges on the flagellar apparatus at a dense plaque just proximal to the distal fiber in motile interphase algal cells (Ringo, 1967; Melkonian 1980; Melkonian & Robenek, 1984; Moestrup; 1978). These microtubules comprise a cruciate "rootlet" system consisting of alternating groups of microtubules in a X-2-X-2 arrangement in the region of the cell closest to the flagellar apparatus (Melkonian, 1984). In *Chlamydomonas* the microtubule rootlets are arranged in a 4-2-4-2 pattern (Fig. 2; Fig. 3C; Ringo, 1967; Goodenough & Weiss, 1978; Weiss, 1984). A finely striated fiber extends from one side of the basal apparatus to the other and is associated with each of the two-membered microtubule rootlets. One of the two-membered rootlets lies below and to one side of the mating structure in mt^+ cells (Goodenough & Weiss, 1978). One of the four-membered rootlets is associated with the eyespot of the cell (Melkonian & Robenek, 1980; 1984; Huang *et al.*, 1982). Melkonian (1980, 1984) has reviewed interactions between flagellar microtubule rootlets and other cell organelles in the green algae. Taken together, these observations have led to the suggestion that microtubule rootlets may provide information for the morphogenesis of cytoplasmic organization and that they may also be involved in signal transduction to and from the flagellar apparatus. Thus, the cortical microtubule-based cytoskeleton in the green algae is illustrative of a role for the microtubule-based cytoskeleton in defining a "coordinate geometry" system within the cytoplasm of eucaryotic cells.

CONTRACTILE FLAGELLAR ROOTS

In addition to the microtubule-based rootlets, algal cells also possess a system of striated fibrous roots (system II roots; Melkonian, 1980). These may be massive striated structures as occur in the prasino-phytes (i.e., *Tetraselmis* or *Platymonas*; Parke & Manton, 1965; Manton & Parke, 1965), or more delicate fibers that link the flagellar apparatus to the nucleus, as in *Chlamydomonas* (Kater, 1929; Katz & McLean, 1979; Salisbury *et al.*, 1988; Salisbury, 1988a). Figure 3 illustrates the localization of centrin in an interphase cell of *Chlamydomonas*. Centrin-based fibers link the two adjacent basal bodies (via the distal fiber) and, in addition, two descending fibers extend from the flagellar apparatus into the cell where they branch to form 8-16 fimbria which embrace the nuclear envelope.

Striated flagellar roots are calcium-sensitive contractile organelles (Salisbury & Floyd, 1978); elevated cytoplasmic Ca^{+2} will induce their contraction. Contractile flagellar roots consist of 5 nm diameter fibers that are composed, in large part, of a 20,000 MW, acidic, phosphoprotein called centrin (Coling & Salisbury, 1987). Centrin accounts for over 60%

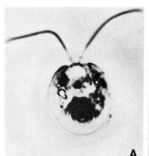

A　　　　　　　　　　B　　　　　　　　C

Figure 3. Centrin distribution in *Chlamydomonas reinhardtii*. A) Phase contrast micrograph. B) Indirect immunofluorescence using monoclonal antibody *17E10* against centrin: bar = 5 μm. C) Diagram (not to scale) illustrating distribution of centrin, four microtubule rootlets, flagellar basal apparatus, and the nucleus. [Reproduced with permission from Salisbury *et al.*, (1988) J Cell Biol 107:635.]

of the total protein of flagellar roots isolated from *Tetraselmis* (Salisbury *et al.*, 1984); therefore, this protein is a major structural protein of the organelle. Recent molecular cloning studies (Huang *et al.*, 1988) have revealed that centrin is a member of the EF hand superfamily (Kretsinger, 1980) of calcium-binding proteins. Contraction of centrin-based fibers, which is induced on calcium-binding, is mediated by a twisting and supercoiling of the 5 nm fibers (Fig. 4; Salisbury, 1983). Biochemical analysis of centrin isotypes and the state of centrin phosphorylation in *Tetraselmis* cells indicate that this protein consists of two isoforms: a dephosphorylated α and a phosphorylated β isotype. Extended flagellar roots show a predominance of phosphorylated β centrin, while contracted flagellar roots show a predominance of dephosphorylated α centrin (Salisbury *et al.*, 1984). These data suggest that centrin is phosphorylated during flagellar root extension and is dephosphorylated during contraction. Studies with detergent extracted *Chlamydomonas* cells indicate that ATP is required to potentiate the system for subsequent calcium-induced contraction (Salisbury *et al.*, 1987).

Nuclear Movement and Flagellar Regeneration

In addition to the roles played by centrin-based systems in basal body reorientation and flagellar excision discussed above, centrin-based flagellar roots function in at least two additional contractile events in the algae: 1) nuclear movement and nuclear shape changes; 2) centrosome/nuclear dynamics during mitosis.

Following flagellar excision in *Chlamydomonas*, individual cells will rapidly regenerate new flagella when returned to favorable conditions.

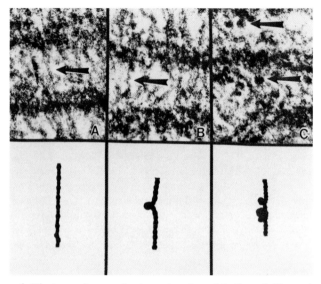

Figure 4. Electron micrographs (upper) and models (lower) illustrating the reorganization of centrin-based filaments during contraction: filaments (A) acquire a bend, twist, or kink (B) and finally supercoil into twists (C) on calcium-binding. [Reprinted with permission from Salisbury (1983), J Submicrosc Cytol 15:105.]

Flagellar roots of *Chlamydomonas* undergo a dramatic and rapid calcium-induced contraction in living cells at the time of flagellar excision (Salisbury *et al.*, 1987). Contraction of flagellar roots results in nuclear shape changes and nuclear movement toward the flagellar basal apparatus. The nucleus remains near the flagellar apparatus for approximately 45 min and subsequently returns to a more central location within the cell; this movement corresponds to the re-extension of the flagellar roots. Figure 5 illustrates the kinetics of flagellar excision and regeneration and nuclear movement in *Chlamydomonas*. Because flagellar root contraction and nuclear movement require the presence of Ca^{2+} in the extracellular milieu, we suspect that a local transient flux of Ca^{2+} into the cell is signaling the contraction. Thus, nuclear position may serve as an indicator of cytoplasmic free calcium levels; nuclear movement may signify a transient rise in cytoplasmic free Ca^{2+} at the time of flagellar excision.

Flagellar regeneration utilizes both a preexisting pool of flagellar precursors and newly synthesized flagellar proteins (Lefebvre & Rosenbaum, 1986). Both the timing of flagellar root contraction following flagellar excision and the kinetics of the onset of flagellar root extension coincide with the timing of induction of flagellar precursor genes, including those for tubulin (Weeks & Collis, 1976). Perhaps the

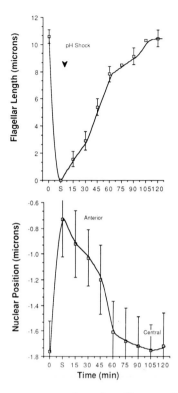

Figure 5. Flagellar regeneration and nuclear movement in *Chlamydomonas* after pH shock induced flagellar excision. A) Kinetics of flagellar regeneration. B) Nuclear position. (Note the change in scale). Initially the nucleus is in a central position, ~ 1.5 μm from the flagellar apparatus. Immediately following pH shock, the nucleus moves to the anterior-most region of the cell where it remains for ~30 to 45 min. The nucleus then returns to a more central position. Each point represents the mean value for 30 individual cells and standard error bars are indicated.

two events are functionally unrelated; flagellar root contraction and flagellar precursor induction may be independent responses to elevated intracellular Ca^{2+} brought about by the conditions used to deflagellate the cells. However, we are led to question whether flagellar root contraction may play a more direct role in the signaling of nuclear events related to the induction and/or transport of flagellar precursors. One possibility is that a physical (centrin-based cytoskeletal) linkage between the flagellar apparatus and the nucleus may mediate the propagation of a signal which results in the induction of flagellar precursor genes. In its simplest sense, we envisage this signal to be a physical or tactile alteration (mediated by centrin contraction) of nuclear matrix and/or chromatin structure which results in the exposure or unmasking of specific regions of DNA, thereby making this DNA accessible to the action of transcriptional machinery. A second possibility is that flagellar root contraction may facilitate the transport of flagellar precursors to the base of the flagellar apparatus. Either of these possibilities may include interactions among the centrin-based cytoskeleton, the nuclear pore complexes, the nuclear lamina and associated transcriptional machinery in a manner consistent with the "gene gating hypothesis" of Blobel (1985). *Chlamydomonas* provides a model system (both genetic and structural) which is uniquely suited for testing the hypothesis that reorganization of cytoskeletal structures may in certain cases be involved in the regulation of gene expression (Peters, 1956; Ashall & Puck, 1984; Salisbury *et al.*, 1987) .

Centrin Dynamics During the Cell Cycle

Our studies show pronounced changes from the interphase organization of centrin during mitosis (Fig. 6; Salisbury *et al.*, 1988). Although dramatic changes in centrin distribution occur, the centrin-based fiber system always remains convergent on the region of the basal bodies and maintains its association with the nuclear envelope. At the interphase/preprophase boundary the centrin-based descending fibers and their branches show a conspicuous contraction, which draws the nucleus toward the flagellar apparatus (Salisbury *et al.*, 1988). This contraction is similar to that described above for nuclear movement following flagellar excision; in mitotic cells this nuclear movement coincides with the loss of flagella, subtle cell shape changes, and a burst of tubulin synthesis (Triemer & Brown, 1974; Ares & Howell, 1982; Piperno & Luck, 1977). The preprophase contraction is transient and is followed in prophase by division and separation of the centrin-based cytoskeleton; the timing of basal body separation is coincident with this event (Coss, 1974). Prior to metaphase the two newly formed centrin foci migrate toward opposite poles of the spindle. By metaphase, centrin has reformed an array of fibers that extend from the poles and delineate the mitotic spindle. At the metaphase/anaphase boundary a second transient contraction of centrin occurs; the timing of chromosome separation is coincident with this event. During telophase the contracted centrin fibers re-extend; thus two distinct half-spindles are delineated. By the time of cytokinesis the two daughter nuclei have reestablished an interphase-like organization of centrin. If the state of centrin-based fiber contraction can be used as a free calcium indicator as suggested above, our observations indicate that there are at least two transient elevations of cytoplasmic free Ca^{2+} in *Chlamydomonas* cells during mitosis; the first occurs at preprophase and the second at the metaphase/anaphase transition. Indeed studies of both plant and animal cells which are amenable to more direct measurements of cytoplasmic free Ca^{2+} levels have demonstrated transient calcium fluxes at the time of the metaphase/anaphase transition (Keith *et al.*, 1985; Poenie *et al.*, 1986; Ratan *et al.*, 1986).

It is possible that the centrin-containing structures of mitotic cells are responsible for basal body/centriole segregation. Jarvik and co-workers (Kuchka & Jarvik, 1982; Wright *et al.*, 1985) have studied a *Chlamydomonas reinhardtii* mutant, *vfl-2*, which has defects in centrin content and localization. Cultures of *vfl-2* contain cells with variable number of flagella; these cells appear to have a defect which results in the mispositioning of basal bodies during cell division thereby giving rise to

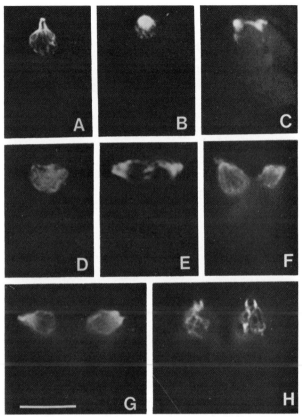

Figure 6. Indirect immunofluorescence using monoclonal antibody *17E10* against centrin illustrating the dynamic behavior of the centrin-based cyto-skeleton in *Chlamydomonas reinhardtii* during the cell cycle. A:) Interphase; B) Preprophase; C) Prophase; D) Metaphase; E) Metaphase/anaphase transition; F-G) Telophase; H) Cytokinesis: bar = 5 μm. [Reprinted with permission from Salisbury *et al.*, (1988), J Cell Biol 107:635.]

daughter cells with inappropriate numbers of basal bodies and subsequently of flagella.

The essential features of centrin dynamics described here for *Chlamydomonas* mitosis are also common to other mitotic cells that we have studied, including mammalian cells (Baron & Salisbury, 1988). It is conceivable that centrin-based contractile phenomena may be in part responsible for the poleward movement of chromosomes during anaphase (cf., Pickett-Heaps *et al.*, 1982).

CONCLUSIONS

The structural basis for the organization of the cytoplasm is determined, in large part, by the cytoskeleton. Algal cells have proven to be useful model systems for studying aspects of cytoskeletal organization and genetics. Centrin-based cytoskeletal structures were first discovered in algal cells. These may be of fundamental importance in the origin and maintenance of cell form because centrin-based motile phenomena have as a common feature a role in the positioning and orientation of the principal microtubule organizing centers of the cell: the flagellar basal apparatus and the centrosome. In addition, centrin appears to be involved in the dynamic behavior of certain microtubules and perhaps in the regulation of nuclear functions, including gene induction and mitosis.

ACKNOWLEDGEMENT

I wish to thank the members of the laboratory for many valuable discussions and comments on this work, and Dr. J. Stein-Taylor and the reviewers of the manuscript for their helpful suggestions. Supported by a grant from NIH (GM35258).

REFERENCES

Aitchison WA, Brown DL (1986). Duplication of the flagellar apparatus and cytoskeletal microtubule system in the alga *Polytomella*. Cell Motil Cytoskel 6:122-127.

Ares M Jr, Howell SH (1982). Cell cycle stage-specific accumulation of mRNAs encoding tubulin and other polypeptides in *Chlamydomonas*. Proc Natl Acad Sci USA 79:5577-5581.

Ashall F, Puck TT (1984). Cytoskeletal involvement in cAMP-induced sensitization of chromatin to nuclease digestion in transformed Chinese hamster ovary K1 cells. Proc Natl Acad Sci USA 81:5145-5149.

Baron A, Salisbury (1988). Identification and localization of a novel, cytoskeletal, centrosome-associated protein in PtK2 cells. J Cell Biol 107: (in press).

Blobel G (1985). Gene gating: A hypothesis. Proc Natl Acad Sci USA 82:8527-8529.

Blum JJ (1971). Existence of a breaking point in cilia and flagella. J Theor Biol 33:257-263.

Brawley SH, Robinson KR (1985). Cytochalasin treatment disrupts the endogenous currents associated with cell polarization in fucoid zygotes: Studies of the role of f-actin in embryogenesis. J Cell Biol 100: 1173-1184.

Brücke E (1861). Die Elementarorganismen: Wiener Sitzber 44 p 386. See Wilson EB (1925). Cited below, p 670 for translation.

Cavalier-Smith T (1974). Basal body and flagellar development during vegetative cell cycle and the sexual cycle of *Chlamydomonas reinhardtii*. J Cell Sci 16:529-556.

Coling DE, Salisbury JL (1987). Purification and characterization of centrin, a novel calcium-modulated contractile protein. J Cell Biol 105:205a.

Coss RA (1974). Mitosis in *Chlamydomonas reinhardtii*: basal bodies and the mitotic apparatus. J Cell Biol 63:325-329.

Cox ER (ed) (1980). "Phytoflagellates." Amsterdam: Elservier North-Holland.

Dawson P, Hulme J, Lloyd C (1985). Monoclonal antibody to intermediate filament antigen cross reacts with higher plant cells. J Cell Biol 100:1793-1798.

Detmers PA, Carboni J, Condeelis JS (1985). Localization of actin in *Chlamydomonas* using antiactin and NBD-phallacidin. Cell Motil 5:415-430.

Dutcher SK (1989). Linkage group XIX in *Chlamydomonas reinhardtii* (Chlorophyceae): Genetic analysis of basal body function and assembly. In Coleman AW, Goff LJ, Stein-Taylor R (eds): "Algae as Experimental Systems", New York: Alan R. Liss Inc, pp 39-53

Fawcett DW, Porter KR (1954). A study of the fine structure of ciliated epithelia. J Morph 94:221-282.

Gaffal KP (1988). The basal body-root complex of *Chlamydomonas reinhardtii* during mitosis. Protoplasma 143:118-129.

Goodenough UW, Adair SW (1989). Recognition proteins of *Chlamydomonas reinhardtii* (Chlorophyceae). In Coleman AW, Goff LJ, Stein-Taylor R (eds): "Algae as Experimental Systems", New York: Alan R. Liss Inc, pp 171-185

Goodenough UW, Weiss RL (1978). Interrelationships between micro-tubules, a striated fiber, and the gametic mating structure of *Chlamydomonas reinhardtii*. J Cell Biol 76:430-438.

Gould RR (1975). The basal bodies of *Chlamydomonas reinhardtii*: Formation from probasal bodies, isolation, and partial characterization. J Cell Biol 65:65-74.

Hoops HJ, Witman G (1983). Outer doublet heterogeneity reveals structural polarity related to beat direction in *Chlamydomonas* flagella. J Cell Biol 97:902-908.

Huang B (1986). *Chlamydomonas reinhardtii*; a model system for the genetic analysis of flagellar structure and motility. Int Rev Cytol 99:181-215.

Huang B, Ramanis Z, Dutcher SK, Luck DJL (1982). Uniflagellar mutants of *Chlamydomonas*: Evidence for the role of basal bodies in transmission of positional information. Cell 29:745-753.

Huang B, Mengerson A, Lee VD (1988). Molecular cloning of a cDNA for centrin, a basal body-associated Ca^{2+}-binding protein: Homology in its protein sequence with calmodulin and the yeast CDC 31 gene product. J Cell Biol 107:133-140.

Huber ME, Wright WG, Lewin RA (1986). Divalent cations and flagellar autotomy in *Chlamydomonas reinhardtii* (Volvocales, Chlorophyta). Phycologia 25:408-411.

Hyams JS, Borisy GG (1975). Flagellar coordination in *Chlamydomonas reinhardtii*: Isolation and reactivation of the flagellar apparatus. Science 189:891-893.

Hyams JS, Borisy GG (1978). Isolated flagellar apparatus of *Chlamydomonas*: Characterization of foward swimming and alteration of waveform and reversal by calcium *in vitro*. J Cell Sci 33:235-253.

Kater JM (1929). Morphology and division of *Chlamydomonas* with reference to the phylogeny of the flagellar neuromotor system. Univ Calif Publ Zool 133:125-168.

Katz KR, McLean RJ (1979). Rhizoplast and rootlet system of the flagellar apparatus of *Chlamydomonas moewusii*. J Cell Sci 39:373-381.

Kamiya R, Witman GB (1984). Submicromolar levels of calcium control the balance of beating between the two flagella in demembranated models of *Chlamydomonas* . J Cell Biol 98:97-107.

Keith CH, Ratan R, Maxfield FR, Bajer A, Shelanski ML (1985). Local cytoplasmic calcium gradients in living cells. Nature 316:848-850.

Kretsinger RH (1980). Structure and evolution of calcium-modulated proteins. Crit Rev Biochem 8:119-174.

Kuchka MR, Jarvik JW (1982). Analysis of flagellar size control using a mutant of *Chlamydomonas reinhardtii* with a variable number of flagella. J Cell Biol 92:170-175.

LaClaire JW II (1989). The algal cytoskeleton I: The interaction of actin and myosin in cytoplasmic movement. In Coleman AW, Goff LJ, Stein-Taylor R (eds): "Algae as Experimental Systems", New York: Alan R. Liss Inc, pp 55--70

Lang NJ (1963). An additional ultrastructural component of flagella. J Cell Biol 19:631-634.

Lefebvre PA, Rosenbaum JL (1986). Regulation of the synthesis and assembly of ciliary and flagellar proteins during regeneration. Annu Rev Cell Biol 2:517-546.

Lewin RA, Burrascano C (1983). Another new kind of *Chlamydomonas* mutant with impaired flagellar autotomy. Experientia 39:1397-1398.

Lewin RA, Lee T-H, Fang L-S (1982). Effects of various agents on flagellar activity, flagellar autotomy and cell viability in four species of *Chlamydomonas* (Chlorophyta: Volvocales). In Amos WB, Duckett JG (eds): "Prokaryotic and Eukaryotic Flagella" Cambridge: Cambridge University Press, pp 421-437.

Lewin RA, Lee KW (1985). Autotomy of algal flagella: Electron microscope studies of *Chlamydomonas* (Chlorophyceae) and *Tetraselmis* (Prasinophyceae). Phycologia 24:311-316.

Manton I (1956). Plant cilia and associated organelles. In Rudnick (ed): "Cellular Mechanisms in Differentiation and Growth" Princeton NJ: Princeton University Press, pp 61-72.

Manton I, Parke M (1965). Observations on the fine structure of two species of *Platymonas* with special reference to flagellar scales and the mode of origin of the theca. J Mar Biol Ass UK 45:743-754.

Mazia D (1984) Centrosomes and mitotic poles. Exp Cell Res 153:1-15.

McFadden GI, Schulze D, Surek B, Salisbury JL, Melkonian M (1987). Basal body reorientation mediated by a Ca^{+2}-modulated contractile protein. J Cell Biol 105:903-912.

Melkonian M (1980). Ultrastructural aspects of basal body associated fibrous structures in green algae: A critical review. BioSystems 12:85-104.

Melkonian M (1982). Structural and evolutionary aspects of the flagellar apparatus in green algae and land plants. Taxon 31:255-265.

Melkonian M (1984). Flagellar root-mediated interactions between the flagellar apparatus and cell organelles in green algae. In Wiessner W, Robinson D, Starr RC (eds): "Compartments in Algal Cells and Their Interaction" Berlin: Springer-Verlag pp 96-108.

Melkonian M, Reize IB, Preisig HR (1987). Maturation of a flagellum/ basal body requires more than one cell cycle in algal flagellates: studies on*Nephroselmus olvacea* (Prasinophyceae). In Wiessner W, Robinson DG, Starr RC (eds): "Algal Development; Molecular and Cellular Aspects." Berlin: Springer-Verlag pp 102-113.

Melkonian M, Robenek H (1980). Eyespot membranes of *Chlamydomonas reinhardtii*: Freeze-fracture study. J Ultrastr Res 72:90-102.

Melkonian M, Robenek H (1984). The eyespot apparatus of flagellated green algae: A critical review. In Round FE, Chapman R (eds): "Progress in Phycological Research" Bristol, UK: Bio Press Ltd vol 3 pp193-268.

Miller CCJ, Duckett JG, Downs MJ, Cowell I, Dowding AJ, Virtanen I, Anderton BH (1985). Plant cytoskeletons contain intermediate filament-related proteins. Biochem Soc Trans 13:960-961.

Moestrup Ø (1978). On the phylogenetic validity of the flagellar apparatus in green algae and other chlorophyll a and b containing plants. Biosystems 10:117-144.

Numata O, Sugai T, Watanabe Y (1985). Control of germ cell nuclear behavior at fertilization by *Tetrahymena* intermediate filament protein. Nature 314:192-194.

Omoto CK, Brokaw CJ (1985). Bending patterns of *Chlamydomonas* flagella II: Calcium effects on reactivated *Chlamydomonas* flagella. Cell Motil 5:53-60.

Parke M, Manton I (1965). Preliminary observations on the fine structure of *Prasinocladus marinus*. J Mar Biol Ass UK 45:525-536.

Peters RA (1956). Hormones and the cytoskeleton. Nature 177:426.

Pickett-Heaps JD (1975). "Green Algae: Structure, Reproduction, and Evolution in Selected Genera." Sunderland, MA: Sinauer Assoc.

Pickett-Heaps JD (1986). Mitotic mechanisms: an alternative view. Trends Biochem Sci 11:504-507.

Pickett-Heaps JD, Tippit DH, Porter KR (1982). Rethinking mitosis. Cell 29:729-744.

Piperno G, Luck DJL (1977). Microtubular proteins of *Chlamydomonas reinhardtii*: an immunochemical study based on the use of an antibody specific for the beta tubulin subunit. J Biol Chem 252:383-391.

Poenie M, Alderton J, Steinhardt R, Tsien R (1986). Calcium rises abruptly and briefly throughout the cell at the onset of anaphase. Science 233:886-889.

Porter KR (1984). The cytomatrix: A short history of its study. J Cell Biol 99(1:2):3s-12s.

Quatrano RS, Kropf DL (1989). Polarization in *Fucus* (Phaeophyceae) zygotes: Investigations of Ca^{2+}, microfilaments and the cell wall. In Coleman AW, Goff LJ, Stein-Taylor R (eds): "Algae as Experimental Systems", New York: Alan R. Liss Inc, pp 111-119

Ratan RR, Shelanski ML, Maxfield FR (1986) Transition from metaphase to anaphase is accompanied by local changes in cytoplasmic free calcium in PtK2 kidney epithelial cells. Proc Natl Acad Sci USA 83:5136-5140.

Ringo DL (1967). Flagellar motion and fine structure of the flagellar apparatus in *Chlamydomonas*. J Cell Biol 33:543-571.

Sale WS, Satir P (1977). Direction of active sliding of microtubules in *Tetrahymena* cilia. Proc Natl Acad Sci USA 74:2045-2029.

Salisbury JL (1983). Contractile flagellar roots: The role of calcium. J Submicrosc Cytol 15:105-110.

Salisbury JL (1988). The lost neuromotor apparatus of *Chlamydomonas*: Rediscovered. J Protozoology 35:574-577.

Salisbury JL, Floyd G (1978). Calcium-induced contraction of the rhizoplast of a quadri-flagellate green alga. Science 202:975-978.

Salisbury JL, Swanson JA, Floyd GL, Hall R, Maihle N (1981). Ultrastructure of the flagellar apparatus of the green alga *Tetraselmis subcordiformis*: With special consideration given to the function of the rhizoplast and rhizanchora. Protoplasma 107:1-11.

Salisbury JL, Baron A, Surek B, Melkonian M (1984). Striated flagellar roots: Isolation and partial characterization of a calcium-modulated contractile organelle. J Cell Biol 99:962-970.

Salisbury JL, Baron A, Coling D, Martindale V, Sanders M (1986). Calcium-modulated contractile proteins associated with the eucaryotic centrosome. Cell Motility and the Cytoskel 6:193-197.

Salisbury JL, Sanders M, Harpst L (1987). Flagellar root contraction and nuclear movement during flagellar regeneration in *Chlamydomonas reinhardtii*. J Cell Biol 105:1799-1805.

Salisbury JL, Baron AT, Sanders M (1988). The centrin-based cytoskeleton of *Chlamydomonas reinhardtii*: Distribution in interphase and mitotic cells. J Cell Biol 107:635-641.

Satir P (1982). Mechnanisms and controls of microtubule sliding in cilia. Symp Soc Exp Biol 35:179-201.

Satir P, Ojakian GK (1979). Plant cilia. In Haupt W, Feinleib ME (eds): "Encyclopedia of Plant Physiology New Series Vol 7 Physiology of Movements." Berlin: Springer-Verlag, pp 224-249.

Schulze D, Robenek H, McFadden GI, Melkonian M (1987). Immunolocalization of a Ca^{2+}-modulated contractile protein in the flagellar apparatus of green algae: the nucleus-basal body connector. Eur J Cell Biol 45:51-61.

Triemer RE, Brown RM (1974). Cell division in *Chlamydomonas moewusii*. J Phycol 10:419-433.

Van Beneden E, (1883). Recherches sur la maturation de l'oeuf, la fecondation et la division cellulaire. Arch Biol 4:265-641.

vanden Ende H, Tomson AM, Demets R, Kooijman R (1989). Modulation of sexual agglutinability in *Chlamydomonas eugametos*. (Chlorophyceae). In Coleman AW, Goff LJ, Stein-Taylor R (eds): "Algae as Experimental Systems", New York: Alan R. Liss Inc, pp 187-200

Weeks DP, Collis PS (1976). Induction of microtubule protein synthesis in *Chlamydomonas reinhardtii* during flagellar regeneration. Cell 9:15-27.

Weiss RL (1984). Ultrastructure of the flagellar roots in *Chlamydomonas* gametes. J Cell Sci 67:133-143.

Wheatley DN (1982). "The Centriole: A Central Enigma in Cell Biology." Amsterdam: Elsevier Biomedical.

Wilson EB (1925). "The Cell in Development and Heredity." 3rd ed. New York: Macmillan.

Witman GB, Carlson K, Berliner J, Rosenbaum JL (1972). *Chlamydomonas* flagella I. Isolation and electrophoretic analysis of microtubules, matrix, membranes, and mastigonemes. J Cell Biol 54:507-539.

Wright RL, Salisbury JL, Jarvik J (1985). A nucleus-basal body connector in *Chlamydomonas reinhardtii* that may function in basal body segregation. J Cell Biol 101:1903-1912.

Algae as Experimental Systems pages 39–53
© 1989 Alan R. Liss, Inc.

LINKAGE GROUP XIX IN *CHLAMYDOMONAS REINHARDTII*
(CHLOROPHYCEAE): GENETIC ANALYSIS OF BASAL BODY
FUNCTION AND ASSEMBLY

Susan K. Dutcher
Department of Molecular, Cellular,
and Developmental Biology
University of Colorado
Boulder, Colorado 80309-0347

The unicellular, green alga *Chlamydomonas reinhardtii* is used
extensively as a model system to study flagellar assembly and function
(Randall & Starling, 1972; Luck, 1983). The flagella are located at
the anterior end of the cell and can propel the cell in a forward direction by a breast-like stroke or in a backwards direction by a sinusoidal
stroke (Gibbons, 1981). At the proximal end of each flagellum is the
basal body, whose nine triplet microtubules serve as the template for
the assembly of the nine doublet microtubules of the flagellar axoneme
(Fawcett & Porter, 1954; Ringo, 1967) (Fig. 1).

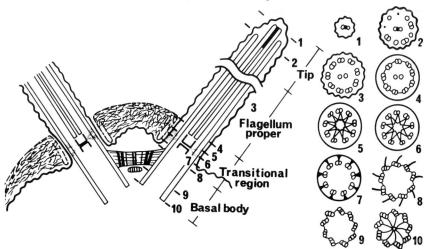

Figure 1: Diagram of morphology of basal bodies and flagella of
Chlamydomonas reinhardtii. (Reproduced by permission of Sinauer
Associated, Inc. and Dr. J. Pickett-Heaps)

By two-dimensional gel electrophoresis, there are at least 250 polypeptide components in addition to α- and β-tubulins, which comprise the majority of the protein mass in preparations enriched in basal bodies (Dutcher, 1986). In flagellar axoneme preparations, 200 polypeptide components are observed (Luck, 1983; Huang, 1986) and only thirty of the polypeptides in the two structures comigrate (Dutcher, 1986). A genetic approach to understanding the basal body and flagellum has been undertaken because of their morphological and biochemical complexity.

Cells that lack flagella or that assemble aberrant flagella are unable to oppose gravity and thus sink when grown in liquid medium (Fig. 2). The flagella are also required for the recognition of cells of the opposite mating-type, which initiates the mating response. Using these phenotypes, various screens have been used to isolate a large number of mutations that affect flagellar function and assembly (Randall & Starling, 1972; Luck, 1983; Goodenough & St. Clair, 1975). Over 240 independent mutant alleles have been isolated.

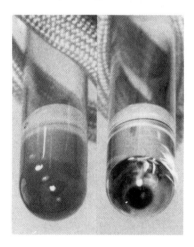

Figure 2: Chlamydomonas cultures in liquid medium. Cells on left are wild-type and swim to air-liquid interface; cells on right contain a flagellar assembly mutations and form a pellet.

Chlamydomonas is generally propagated as a haploid organism. However, cells of opposite mating-type can be induced by starvation to undergo pairwise conjugation and become dormant zygotes (Smith & Regnery, 1950; Sager & Granick, 1953) These zygotes, upon the proper nutritional and environmental signals, initiate meiosis and produce four meiotic products that can be dissected and analyzed. A small percentage of newly conjugated cells skip meiosis and become mitotically growing diploid cells (Ebersold, 1967), which reorganize their cytoskeleton. In most respects, they resemble haploid cells. They can be either propagated by mitotic cell divisions or induced to conjugate with haploid cells. Triploid zygotes complete meiosis and normally produce aneuploid progeny. By blocking nuclear fusion between the haploid and diploid nuclei with microtubule antagonists, four haploid progeny from the diploid nucleus can be recovered. This genetic trick allows for the effective sporulation of otherwise permanent diploids (Dutcher, 1988a). Thus, a variety of different types of genetic analyses are possible in *Chlamydomonas*.

Mutations that define at least 42 loci have been identified that affect the function of the flagella. Mutations in this class assemble flagella, but these flagella are either non-functional or function abnormally (Luck, 1983). Many of these mutations affect the assembly of specific axonemal substructures such as the central pair microtubules (Witman *et al.*, 1978; Adams *et al.*, 1981; Dutcher *et al.*, 1984) the radial spokes (Witman *et al.*, 1978; Luck *et al.*, 1977; Huang *et al.*, 1981); the outer and inner dynein arms (Huang *et al.*, 1979; 1982a; Kamiya & Witman, 1984; Mitchell & Rosenbaum, 1985) or doublet projections (Segal *et al.*, 1984). A second class of flagellar mutations, which is defined by at least 33 additional loci, alter the assembly of flagella. In this class, there are mutations that assemble either no flagella (Huang *et al.*, 1977; Adams *et al.*, 1982; Dutcher, 1986), abnormal length flagella that can be either long or short, (McVittie, 1972; Kutcha & Jarvik, 1987; Barsel *et al.*, 1988), or an abnormal number of flagella (Adams *et al.*, 1985; Huang *et al.*, 1982b; Wright *et al.*, 1983). The genetic map positions of flagellar mutations is shown in Fig. 3. Loci identified by phenotypes affecting flagellar function map to 15 of the 19 linkage groups. Loci identified by phenotypes affecting flagellar assembly map to 9 of the 19 linkage groups. Strikingly, a large number of these loci are located on a single linkage group (linkage group XIX or the *uni* linkage group) (Huang *et al.*, 1982b; Dutcher, 1986; Ramanis & Luck, 1986). Linkage group XIX has a number of unusual properties that distinguish it from other linkage groups in both *Chlamydomonas* and other eukaryotes. First, all of the loci that have been mapped to linkage group XIX affect flagellar assembly and function. Second, the genetic map of linkage group XIX is circular. Third , it shows altered recombinational properties.

CLUSTERING OF RELATED GENES

All the mutations mapped to date on linkage group XIX appear to affect processes involving microtubules. Loci can be classified into seven different classes based on their mutant phenotypes. Some of these classes may overlap. Mutants in class I fail to assemble flagella; they define 8 loci. Most of the loci are represented by only a single mutant allele and the alleles are temperature-sensitive; the strains assemble flagella at 21°C, but fail to assemble flagella at 32°C. Multiple alleles at the *fla10* locus have been isolated and examined. Nine alleles show a conditional temperature-sensitive phenotype, while four alleles show an aflagellate phenotype at all temperatures and

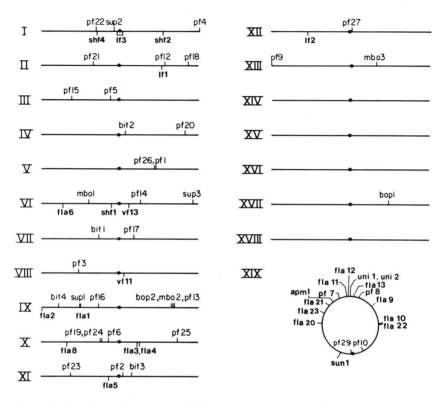

Figure 3: Genetic map of *Chlamydomonas reinhardtii* showing location of loci that affect flagellar function (above the line) and location of loci that affect flagellar assembly (below in the line in bold-face type). Mutations at the *enh1* and *sun5* loci map within 1 cM of *uni1*, but order not determined. The mutations at the *apm9* locus maps with 2 cM of *fla9* but order is not determined.

conditions tested (Dutcher, 1988b). Mutants in class II assemble only one of the two flagella and are known by the acronym *uni*. The assembled flagellum is always found in a specific orientation with respect to the eyespot, which is an asymmetrically positioned organelle involved in phototaxis (Huang *et al.*, 1982b). These mutant pheno-types indicate differences between the two basal bodies and their flagella and a non-random positioning of the two basal bodies in the cell. Mutants in class III assemble short flagella and define three loci (Dutcher, 1986; 1988b). Mutants in class IV assemble flagella that function abnormally; they define two loci. One of these mutations, *pf10*, is a conditional allele; the restrictive condition is the absence of light and the permissive condition is light intensity of greater than 4,000 ergs/cm^2·sec (Dutcher *et al.*, 1988). Mutants in class V are identified as suppressors or enhancers of the Uni phenotype (Dutcher, 1986; Ramanis & Luck, 1986). The suppressor mutations have no detectable phenotype besides this suppression of the uniflagellar phenotype. The one enhancer mutation displays a weak temperature-sensitive flagellar assembly phenotype as well as making *uni* strains aflagellate. Mutants in class VI confer resistance to several herbicides that interact with β-tubulin (Hess and Bayer, 1977) and inhibit cell division. This class includes two loci, *apm1* (James *et al.*, 1988) and *apm3* (S.K. Dutcher, unpublished work). It has been deduced from restriction fragment length polymorphism data that these loci do not encode tubulin genes (James *et al.*, 1988). The mechanism by which these loci afford resistance to the herbicide is not known. Alleles at *apm1* show both cold- and temperature-sensitive lethal phenotypes (James *et al.*, 1988; S.K. Dutcher, unpublished work). Class VII is defined by a single mutant allele, *fla10-5*. This *fla10* allele displays two distinct phenotypes. It is temperature-sensitive for flagellar assembly and has an increased cell division cycle time. The doubling time is 29 h instead of 8 h for the parental strain at 21 ° C. This allele fails to complement other *fla10* alleles for the flagellar defect. However, three other *fla10* alleles all complement the slow cell division cycle time. These two phenotypes cosegregate and corevert; which suggests the phenotypes result from the same mutation (Dutcher, 1988b). In summary, these last two classes suggest that some products of linkage group XIX may participate in processes besides flagellar assembly and function.

The current hypothesis is that these linkage group XIX mutations may all define functions needed for the assembly of the basal body of *Chlamydomonas* and that linkage group XIX may encode only gene products needed in this structure. It is clear that genes on other linkage groups are required for flagellar and basal body assembly; these include the *fa* locus (Salisbury, this volume) and *bald-2* (Goodenough and St. Clair, 1975; S.K. Dutcher, unpublished work). The hypothesis that linkage group XIX gene products are used in basal body assembly is based on two pieces of evidence. The first is that aberrant basal body complexes are observed in the *uni* mutants by thin-section electron microscopy (Huang *et al.*, 1982b). The second is the behavior of many of these mutations in dikaryotic zygotes (Dutcher, 1986). In newly formed zygotes, the flagellar apparatuses of the gametic cells remain separate, and complementation of flagellar defects can be assayed soon after cell fusion, but before nuclear fusion. Within the zygotic cell, there is a large pool of polypeptides that can be used in flagellar assembly. Complementation of paralyzed or abnormal flagella can be easily ascertained in the presence of protein synthesis inhibitors. For many of the mutations that have nonfunctional flagella or defects in length control, the defect can be rescued in the absence of protein synthesis in matings between a mutant and a wild-type cell. Thus, for these mutations, the mutant phenotype is recessive and shows *in situ* rescue (Table 1). For a small subset of mutations, rescue can be achieved following amputation of the flagella and regrowth in the presence of wild-type cytoplasm. For these mutations, the mutant phenotype is recessive but requires regeneration of the flagella in the mixed cytoplasm. For five mutations on linkage group XIX that were tested, the mutant phenotype can not be rescued in the zygote by wild-type cytoplasm under any of these conditions although the mutations are recessive when tested in heterozygous vegetatively dividing diploids. For example, diploid cells heterozygous for the *uni1* mutation have two flagella and are motile, therefore the mutation is recessive to the wild-type allele (Huang *et al.*, 1982b; Dutcher, 1986). When newly formed heterozygous zygotes are examined, they have three flagella. Presumably, two of the flagella are from the wild-type parent and one is from the mutant parent. No rescue is observed *in situ*. When the flagella are amputated, only three flagella regenerate. So whereas the *uni* mutation is recessive in diploid cells, in dikaryons the aflagellate basal body remains aflagellate and the mutation is said to be autonomous (Table 1). Based on these experiments, it seems probable that the primary defect is not in the flagellar axoneme but elsewhere within the cell. A likely location for the primary defect is within the basal body because of its role in flagellar assembly.

Table 1. Behavior of flagellar mutations in
Chlamydomonas reinhardtii dikaryons

Type of rescue	Loci tested	Reference
In situ	*pf1, pf14*	Luck *et al.*, 1977
	pf16, pf19	Starling and Randall, 1971
	pf6	Dutcher *et al.*, 1984
	pf17, pf24, pf25, pf26, pf27	Huang *et al.*, 1981
	sup4	Huang *et al.*, 1982a
	lf1, lf2, lf3	Barsel *et al.*, 1988
	shf1, shf2, shf3	Kutcha and Jarvik, 1987
Regeneration	*pf18*	Starling and Randall, 1971
	fla10	Dutcher, 1986
	sup1, sup2, sup3	Huang *et al.*, 1982a
Autonomous	*uni1, uni2, fla9, pf10, sun1*	Dutcher, 1986

CIRCULARITY OF LINKAGE GROUP XIX

When mapping mutations located on linkage group XIX with respect to the *pf10* mutation, a linear order of loci is obtained outward in both directions. When loci at the ends of these two linear arrays are crossed to one another, these loci appear to be tightly linked to one another rather than distantly linked as would be expected for a linear map. This scenario is true no matter what locus is used as the starting point for the analysis. These data are most consistent with a circular genetic map. In addition to 19 mutant loci, two types of cis-acting sites can be mapped to this linkage group. One is a region that behaves as a centromere with respect to the segregation of other mapped centromeres in *Chlamydomonas*. The behavior of the centromere is consistent with reductional segregation of linkage group XIX at meiosis I; sister chromatids remain together at this division. It maps between *pf10* and *sun1*. The second type of site places boundaries on recombinational interference. There are two of these sites and they map near the *uni1* region and near the centromere. (see below)

A circular genetic map is generated by several different models. The first model is that the genomic nucleic acid molecule is physically circular at the time of meiosis. Because all four products of meiosis are viable and the markers are recovered in a Mendelian pattern in all tetrads analyzed, the distribution of recombination events must produce an even number of exchanges on each chromatid to avoid the production of dimeric molecules. The second model is that the chromosome is physically linear but the number of recombination events on each chromatid is constrained and it is biased toward an even number of exchanges. There can be 0, 2, 4 or more even numbers of events on each chromatid. If there is generally an even number of exchanges, then loci near the ends generally will be parental with respect to one another and appear linked to one another (Stahl, 1967; Dutcher, 1986; Ramanis & Luck, 1986). Third is the possibility that the linkage group is circularly permuted like the T4 genome (Streisinger *et al.*, 1964). If this were true, then some genes would be present in two copies in each spore. This is unlikely because all of the loci show normal Mendelian segregation in crosses with wild-type cells. However, a series of linear, permuted molecules without terminal redundancy could explain the data. Genetically, the first two models are indistinguishable from one another when examining asci with four viable spores if the bias is strong in the second model. The first two models require that the recovered products have an even number of exchanges on each chromatid. When strains heterozygous for three mutations (either *uni1, pf10, fla20* or *uni1, pf10, fla10*) were examined, no exceptions to this prediction were found in about 2600 tetrads with four viable spores for the first cross and in about 600 in the second cross (S.K. Dutcher, unpublished work). If the molecule is circular, this biased pattern of recombination would be required to avoid producing dimeric molecules that would presumably break as they segregate on the mitotic spindle. If the genomic molecule is linear then the rationale for constraining the recombination to an even number of exchanges is obscure; perhaps this pattern of recombination is required for faithful mitotic chromosome segregation.

RECOMBINATION

Recombination on linkage group XIX differs in two other ways from recombination on other linkage groups in *Chlamydomonas*. First, the recombination distances between loci on linkage group XIX is dramatically altered by temperature. Increasing temperature increases the recombinational map distance between pairs of loci and between different loci and the centromere region (Dutcher, 1986; Ramanis & Luck, 1986). The change in map distance can be large; for example, the distance from *pf10* to the centromere is 1.5cM at 16°C and is 9cM at 32°C. Recombination on 12 other linkage groups in *Chlamydomonas* (linkage groups I through XI, XVIII) is not affected by temperature (Dutcher, 1986; Ramanis and Luck, 1986). The temperature-sensitive period occurs at a puzzling time in the life cycle. Recombination occurs by all criteria during pachytene of meiosis I; however, the temperature-sensitive period for linkage group XIX recombination precedes meiosis by at least four days (Ramanis & Luck, 1986). It is not known if recombination on linkage group XIX is occurring at a different time or whether some precondition for recombination is occurring during this period.

A second set of unusual features of linkage group XIX is the occurrence of recombinational interference and the existence of discrete boundaries to this interference. The linkage group can be divided into two recombinational regions. Within each region, there is strong interference. The interference is calculated from the number of nonparental tetrads (four strand double exchanges) observed experimentally and the number of four strand double exchanges expected based on the measured number of tetratype tetrads (single exchange events) observed experimentally. The number of double exchanges is dramatically reduced compared to the expected value (Snow, 1979) for markers within the region from the centromere to near *uni1* and from *uni1* back to the centromere. However, for pairwise crosses that flank the centromere or the region near *uni1*, the frequency of four strand double exchanges increases with respect to the expected frequency of four strand double exchanges. Therefore, two boundaries are proposed and they lie near the region that contains the centromere and a region near the *uni* mutations. The interference may be either chiasma interference or chromatid interference with a reduction in the number of four strand double exchanges with respect to the other classes (Dutcher, 1986). Interference is observed on some of the other linkage groups; pairwise crosses on linkage group V (Ebersold *et al.*, 1962), VII (Dutcher *et al.*, 1988) VIII (Dutcher & Gibbons, 1988), and XI (Dutcher *et al.*, 1988) show interference values from 0.25 to 0.01 as calculated by the method of Snow (1979). This interference differs in that no boundaries are observed.

In summary, recombination on linkage group XIX may use at least one enzyme or activity that is not used for recombination on other linkage groups based on the temperature-sensitive period and the unusual interference pattern. It is not clear if the time of recombination is different on linkage group XIX or whether some precondition for linkage group XIX is set up before premeiotic DNA synthesis (Chiang & Sueoka, 1967; Cavalier-Smith, 1974) It is likely that the recombinational interference is related to the mechanism by which even numbers of crossover events is enforced.

INTERACTIONS WITH LINKAGE GROUP XIX MUTATIONS

Loci that interact with mutant alleles can be used to understand the role of the wild-type product in the cell. Two loci on linkage group XIX show interesting sets of interactions; these are the *pf10* locus and the *apm1* locus. A single mutation at the *pf10* locus produces cells that assemble flagella, but they beat abnormally (Fig. 4.; Inwood, 1985).

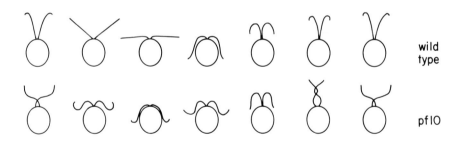

Figure 4: The waveform of wild-type and *pf10* cells traced from video microscopy images. Note the tangling that occurs between the flagella of the *pf10* cell.

Because *pf10* cells sink to form a pellet in liquid medium at low light intensities, suppressors of this phenotype can be obtained easily (Dutcher *et al.*, 1988). These suppressors fall into several unlinked complementation groups; none of these loci map to linkage group XIX. Two of the suppressors, *lis1* and *lis2*, alter the way that cells respond to light signals and thus capitalize on the conditional nature of the *pf10-1* mutation. Alleles at these two loci are recessive to the wild-type allele in heterozygous diploids (Dutcher *et al.*, 1988), but in complementation tests they show another property. A subset of alleles at one *lis* locus are dominant enhancers of a subset of alleles at the other *lis* locus. Dominant enhancement is a genetic interaction in which two recessive mutations at different genes show a mutant phenotype in the double heterozygote. The strain $\dfrac{pf10}{pf10}\ \dfrac{lis1}{LIS1}\ \dfrac{lis2}{LIS2}$ shows suppression of the pf10 phenotype while $\dfrac{pf10}{pf10}\ \dfrac{lis1}{LIS1}$ or $\dfrac{pf10}{pf10}\ \dfrac{lis2}{LIS2}$ are not suppressed. Dominant enhancers of many different mutations have been isolated (Welshons & Von Halle, 1962; Lindsley & Zimm, 1985; Kusch & Edgar, 1986). It is likely that genes that can act as dominant enhancers of one another are sensitive to the dosage of the wild-type gene product. Genes in different pathways (Atkinson, 1985) as well as the same pathway (Matthews & Kaufman, 1987) may show a dominant enhancer phenotype.

The *apm1* mutation is also a dominant enhancer of a second locus, *apm2*, that confers resistance to the herbicide amiprophos-methyl; *apm2* maps to linkage group VIII (James *et al.*, 1988). Each of these mutations is recessive to its wild-type allele and so heterozygous diploids are sensitive to the action of various herbicides. However in the double heterozygote $\dfrac{apm1}{APM1}\ \dfrac{APM2}{apm2}$, the cells are resistant (James *et al.*, 1988). The *apm2* mutation is also temperature-sensitive for growth, but the *apm1* mutation is not a dominant enhancer of this second phenotype. In summary, interactions between various linkage group XIX mutations and with other unlinked mutations, may provide a way to learn what role the basal bodies play in various cellular processes. The genetic interaction between these mutations may indicate an interaction between the wild-type products of these loci.

CONCLUSIONS AND PROSPECTS

Many questions remain about the inheritance and function of linkage group XIX. I am interested in determining what the range of functions encoded by linkage group XIX genes includes. If the hypothesis about the role of linkage group XIX is correct, then this approach provides a convenient and powerful way to study the role of basal bodies and microtubules in the cell. To this end, we are beginning to saturate this linkage group with mutations and to characterize their mutant phenotypes.

Early observations on centrioles in eukaryotic cells have lead to hypotheses about the autonomy of these structures and to the postulate that centrioles/basal bodies may carry their own genetic material

(Lwoff, 1950; Dippell, 1968; Wheatley, 1982). The presence of nucleic acid in centrioles/basal bodies remains hypothetical. Linkage group XIX may be considered a candidate for a centriolar genome. At present, there are no data to support this contention. Experiments with meiotic dikaryons suggest that linkage group XIX segregates as a nuclear associated linkage group and not as a cytoplasmic element (Dutcher, 1988a).

It is clear that there is nothing similar to linkage group XIX in any organism that has been well-studied genetically. Therefore, it is of interest to know why such a chromosome has evolved in *Chlamydomonas reinhardtii*. Currently, it is not known if other species of *Chlamydomonas* have a similar linkage group. Because many of these genes appear to be necessary for flagellar assembly and function, it is reasonable to guess that homologous genes have been incorporated into other chromosomes in these other organisms. One of the other intriguing questions about linkage group XIX is whether it is physically circular or linear. The most unequivocal way to answer this question will be to obtain physical probes for this molecule. These probes will also allow the unequivocal localization of the molecule in *Chlamydomonas* and in other eukaryotes.

ACKNOWLEDGEMENTS

I thank Jeff Holmes, Karla Kirkegaard, Michael Klymkowsky, Ford Lux and Gary Stormo for helpful comments on drafts of this manuscript. This work was supported by a grant from the National Institutes of Health (GM-32843) and an award from the Searle Scholar Program of the Chicago Community Trust.

REFERENCES

Adams GMW, Huang B, Piperno G, Luck DJL (1981). Central pair microtubule complex of Chlamydomonas flagella: Polypeptide composition as revealed by analysis of mutants. J Cell Biol 91: 69-76.

Adams GMW, Huang B, Luck DJL (1982). Temperature-sensitive, assembly defective flagellar mutants of *Chlamydomonas reinhardtii*. Genetics 100: 579-586.

Adams GMW, Wright RL, Jarvik JW (1985). Defective temporal and spatial control of flagellar assembly in a mutant of *Chlamydomonas reinhardtii* with variable flagellar number. J Cell Biol 100: 955-964.

Atkinson KD (1985). Two recessive suppressors of *Saccharomyces cerevisiae cho1* that are unlinked but fall in the same complementation group. Genetics 111: 1-6.

Barsel SE, Wexler DE, Lefebvre PA (1988). Genetic analysis of long-flagella mutants of *Chlamydomonas reinhardtii*. *Genetics*, 118: 637-648.

Cavalier-Smith T. (1974). Basal body and flagellar development during the vegetative cell cycle and the sexual cycle of *Chlamydomonas reinhardtii*. J Cell Sci 16: 529-556.

Chiang KS, Sueoka N (1967). Replication of chromosomal and cytoplasmic DNA during mitosis and meiosis in the eucaryote *Chlamydomonas reinhardi*. J Cell Physiol 70 supl: 89-112.

Dippell RV (1968). The development of basal bodies in Paramecium. Proc Nat Acad Sci USA 61: 461-468.

Dutcher SK (1986). Genetic properties of linkage group XIX in *Chlamydomonas reinhardtii*. In Wickner RB, Hinnebusch A, Lambowitz AM, Gunsalus IC, Hollaender A (eds): "Extrachromosomal Elements in Lower Eukaryotes," New York: Plenum Publishing Corp, pp 303-325.

Dutcher SK (1988a). Nuclear fusion-defective phenocopies in *Chlamydomonas reinhardtii*: Mating-type functions for meiosis can act through the cytoplasm. Proc Nat Acad Sci USA 85: 3946-3951.

Dutcher SK (1988b). Genetic analysis of microtubule organizing centers. In Warner FD, McIntosh JR (eds): "Cell Movement, Vol. II," New York: Alan R Liss, Inc, in press.

Dutcher SK, Gibbons W (1988). Isolation and characterization of dominant tunicamycin resistance mutations in *Chlamydomonas reinhardtii* (Chlorophyceae). J Phycol 24: 230-236.

Dutcher SK, Gibbons W, Inwood WB (1988). A genetic analysis of suppressors of the *PF10* mutation in *Chlamydomonas reinhardtii*. Genetics: in press.

Dutcher SK, Huang B, Luck DJL (1984). Genetic analysis of central pair microtubules of the flagella of *Chlamydomonas reinhardtii*. J Cell Biol 98: 229-236.

Ebersold WT (1967). *Chlamydomonas reinhardtii*: Heterozygous diploid strains. Science (Wash) 157: 446-449.

Ebersold WT, Levine RP, Levine EE, Olmsted MA (1962). Linkage maps in *Chlamydomonas reinhardtii*. Genetics 47: 531-543.

Fawcett DW, Porter KR (1954). A study of the fine structure of ciliated epithelia. J Morphol 94: 221-281.

Gibbons I (1981). Cilia and flagella in eukaryotes. J Cell Biol 91: s107- s124.

Goodenough UW, St. Clair HS (1975). *Bald-2*: a mutation affecting the formation of double and triplet sets of microtubules in *Chlamydomonas reinhardtii*. J Cell Biol 66: 480-491.

Hess FD, Bayer DE (1977). Binding of the herbicide trifluralin to *Chlamydomonas* flagellar tubulin. J Cell Sci 24: 351-360.

Huang BP-H (1986). *Chlamydomonas reinhardtii*: A model system for the genetic analysis of flagellar structure and motility. Int Rev Cytol 99: 181-216.

Huang B, Piperno G, Luck DJL (1979). Paralyzed flagella mutants of *Chlamydomonas reinhardtii* defective for axonemal doublet microtubule arms. J Biol Chem 254: 3091-3099.

Huang B, Piperno G, Ramanis Z, Luck DJL (1981). Radial spokes of Chlamydomonas flagella: Genetic analysis of assembly and function. J Cell Biol 88: 80-88.

Huang B, Ramanis Z, Luck DJL (1982a). Suppressor mutations in Chlamydomonas reveal a regulatory mechanism for flagellar function. Cell 28: 115-124.

Huang B, Ramanis Z, Dutcher SK, Luck DJL (1982b). Uniflagellar mutants of Chlamydomonas: evidence for the role of basal bodies in the transmission of positional information. Cell 29: 745-753.

Huang B, Rifkin MR, Luck DJL (1977). Temperature-sensitive mutations affecting flagellar assembly and function in *Chlamydomonas reinhardtii*. J Cell Biol 72: 67-85.

James S W, Ranum LPW, Silflow C, Lefebvre PA (1988). Mutants resistant to anti-microtubule herbicides map to a locus on the *uni* linkage group in *Chlamydomonas reinhardtii*. Genetics 118: 141-147.

Kamiya R, Witman GB (1984). Submicromolar levels of calcium control the balance of beating between the two flagella in demembranated models of Chlamydomonas. J Cell Biol 98: 97-107.

Kusch M, Edgar RS (1986). Genetic studies of unusual loci that affect body shape of the nematode *Caenorhabditis elegans* and may code for cuticle structural proteins. Genetics 113: 621-639.

Kutcha MR, Jarvik JW (1987). Short-flagella mutants of *Chlamydomonas reinhardtii*. Genetics 115: 685-691.

Lindsley DL, Zimm G (1985). The genome of *Drosophila melanogaster*. Dros Inf Ser 62: 100-103.

Luck DJL (1983). Genetic and biochemical dissection of the eukaryotic flagellum. J Cell Biol 98: 789-794

Luck DJL, Piperno G, Ramanis Z, Huang B (1977). Flagellar mutants of Chlamydomonas: studies of radial spoke-defective strains by dikaryon and revertant analysis. Proc Nat Acad Sci USA 74: 3456-3460.

Lwoff A (1950). "Problems in Morphogenesis in Ciliates: The Kinetosome in Development, Reproduction, and Evolution", New York: John Wiley & Sons.

Matthews KA, Kaufman TC (1987). Developmental consequences of mutations i the 84B α-tubulin gene of *Drosophila melanogaster*. Dev Biol 119: 100-114.

McVittie A (1972). Genetic studies on flagellum mutants of *Chlamydomonas reinhardtii*. Genet Res Camb 19: 157-164.

Mitchell DR, Rosenbaum JL (1985). A motile Chlamydomonas flagellar mutant that lacks outer dynein arms. J Cell Biol 100: 1228-1234.

Ramanis Z, Luck DJL (1986). Loci affecting flagellar assembly and function map to an unusual linkage group in *Chlamydomonas reinhardtii*. Proc Nat Acad Sci USA 83: 423-426.

Randall J, Starling D (1972). Genetic determinants of flagellum phenotype in *Chlamydomonas reinhardtii*. In Beatty RA, Glueksohn-Waelsch S (eds): "Genetics of the Spermatozoon," Copenhagen: Bogtrykkeriet Forum, pp 13-36.

Ringo DL (1967). Flagellar motion and fine structure of the flagellar apparatus in *Chlamydomonas reinhardtii*. J Cell Biol 33: 543-571.

Sager R, Granick S (1953). Nutritional studies with *Chlamydomonas reinhardtii*. Ann NY Acad Sci 56: 831-838.

Salisbury, JL (1989). Algal Centrin: Calcium-sensitive contractile organelles. In Coleman AW, Goff LJ, Stein-Taylor JR (eds): "Algae as Experimental Systems," New York: Alan R Liss, Inc, pp 19-37

Segal RA, Huang B, Ramanis Z, Luck DJL (1984). Mutant strains of *Chlamydomonas reinhardtii* that move backwards only. J Cell Biol 98: 2026-2034.

Smith G, Regnery DG (1950). Inheritance of sexuality in *Chlamydomonas reinhardi*. Proc Nat Acad Sci USA 36: 246-248.

Snow R (1979). Maximum likelihood of linkage and interference from tetrad data. Genetics 92: 231-245.

Stahl F (1967). Circular genetic maps. J Cell Physiol 70 sup 1: 1-12.

Starling D, Randall J (1971). The flagella of temporary dikaryons of *Chlamydomonas reinhardtii*. Genet Res Camb 18: 107-118.

Streisinger G, Edgar RS, Denhardt GH (1964). Chromosome structure in phage T4. Proc Nat Acad Sci USA 51: 775-779.

Welshons WJ, Von Halle ES (1962). Pseudoallelism at the Notch locus in Drosophila. Genetics 47: 743-759.

Wheatley DN (1982). "The Centriole, a Central Enigma in Cell Biology", New York: Elsevier Biomedical Press.

Witman GB, Plummer J, Sander G (1978). Chlamydomonas flagellar mutants lacking radial spokes and central tubules. J Cell Biol 76: 729-747.

Wright RL, Chojnacki B, Jarvik JW (1983). Abnormal basal body number, location, and orientation in a striated fiber defective mutant of *Chlamydomonas reinhardtii*. J Cell Biol 96: 1697-1707.

Algae as Experimental Systems pages 55-70
© 1989 Alan R. Liss, Inc.

THE INTERACTION OF ACTIN AND MYOSIN IN CYTOPLASMIC MOVEMENT

John W. La Claire II

Department of Botany
University of Texas at Austin
Austin, Texas 78713-7640

Movement is a fundamental property of living things, and algal cells provide intriguing and invaluable systems for studying a great variety of motility phenomena. By definition, cell motility processes utilize some form of chemical energy (usually adenosine 5'-triphosphate; ATP), and typically calcium ions play a role in regulating movement - either stimulating or inhibiting it depending on the type of motility. The majority of motile phenomena studied so far appear to involve one (or both) of two major cytoskeletal components: actin-containing microfilaments (MFs) and tubulin-containing microtubules (MTs). The force-generating (mechanochemical) ATPases for each of these systems are myosin and dynein (or kinesin), respectively. The present discussion will be limited primarily to the actin/myosin system and models of its involvement with bulk cytoplasmic movement in algal cells. Tubulin-based movement is covered in other chapters as are oriented movements of individual organelles such as chromosomes, plastids and flagellar components.

GENERAL FEATURES OF ACTIN, MYOSIN AND THEIR INTERACTION

Both actin and myosin are believed to occur in most if not all eukaryotic cells (e.g., Taylor, 1986). Current views of how actin and myosin may effect movement in nonmuscle cells are primarily based on the classical "sliding filament" model of actomyosin functioning in vertebrate striated muscle (see Murray & Weber, 1974). The globular actin molecule, which consists of a single subunit, is

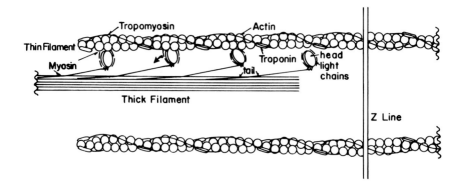

Fig. 1. Schematic diagram of muscle sarcomere portion.
Thin filaments, anchored at "plus" end in Z line, consist of
two helically-wound chains of globular actin molecules,
tropomyosin and troponin. Thick filaments contain myosin
molecules attached at tails, with double heads and light
chains exposed on surface of filament. Force generation
occurs when myosin heads sequentially bind to actin, split
ATP, swivel toward tail (arrow) and release. (After Murray
& Weber, 1974; with permission).

polymerized into double helical "thin filaments" in the
muscle sarcomere (Fig. 1). Actin filaments in general are
polarized at the molecular level, such that net polymerization occurs at the end pointing away from the direction of
its sliding (i.e., "plus end"), and net depolymerization
occurs at the end toward the direction of its movement
(i.e., "minus end"). The myosin molecule is typically a
hexamer, consisting of two heavy chains and two pairs of
light chains. Each pair of heavy chains has a hinged tail
region and a pair of swivelling heads. The tail regions
aggregate myosin molecules into bipolar "thick filaments",
with the head regions (and associated light chains) exposed
on the surface of the thick filament. In the presence of
ATP and micromolar Ca^{2+}, myosin heads bind to available
sites on the thin filaments. Binding stimulates the myosin
heads to split ATP into adenosine diphosphate and inorganic
phosphate, and energy is released. Thus, myosin is an
actin-activated [and Mg-dependent] ATPase. The chemical
energy is transformed into mechanical energy by a conforma-

tional change in myosin, in which the heads swivel toward
the tail region causing a sliding on the thin filaments.
After sliding, the heads release from the thin filaments and
return to their normal conformation. If both ATP and Ca^{2+}
remain available, the myosin heads will bind further down
the thin filament and continue to cycle through these
events, thus effecting contraction. If ATP supplies are
exhausted, the myosin heads remain attached to the thin
filaments, in what are termed "rigor complexes". The ATPase
activity, which swivels the myosin heads to generate force
and movement, and the architecture of the muscle sarcomere
result in muscle shortening (contraction) by an inward sliding of the thin filaments along the myosin thick filaments.

Also according to the model, the Ca^{2+}-stimulation of
muscle contraction results from the troponin-tropomyosin
complex associated with the thin filaments (Fig. 1). In the
absence of Ca^{2+}, myosin-binding sites on the thin filaments
are blocked by tropomyosin. When the Ca^{2+}-sensitive subunit
of troponin binds Ca^{2+}, troponin undergoes a conformational
change that shifts the position of tropomyosin slightly,
thereby exposing the myosin-binding sites. Therefore, regulation of contraction in vertebrate striated muscle is indirectly linked to actin filaments. Other systems may have
Ca^{2+}-sensitivity linked to myosin itself, either directly or
via calmodulin-mediated kinases and phosphatases, as discussed below.

BIOCHEMISTRY OF ALGAL ACTIN AND MYOSIN

Although actin has been detected and/or localized in a
number of algal species (e.g., Menzel & Schliwa, 1986; Palevitz & Hepler, 1975), there has been but one biochemical
study of algal actin. Williamson et al. (1987) found that
bundles of actin filaments in the alga Chara consist of a
single actin isoform, similar in molecular weight (43 kDa)
and isoelectric point (5.5) to other actins. Similarly,
little is known about algal myosin. Nitella (which is
similar to Chara) myosin consists of at least one 200 kDa
heavy chain (light chain composition is unknown), and has
ATPase activity characteristic of many non-muscle myosins
(Kato & Tonomura, 1977; Vorob'eva & Poglazov, 1963). The
paucity of information about algal (or any plant) actin and
myosin precludes any generalizations.

MODELS OF ACTOMYOSIN-BASED CYTOPLASMIC MOVEMENT IN ALGAE

Two forms of bulk cytoplasmic movement are especially
evident in coenocytic and giant-celled green algae: cytoplasmic streaming and wound-induced cytoplasmic contractions. The first recorded observations of streaming were
made with charophytes more than two centuries ago (see Allen
& Allen, 1978). Since that time, most work in this area has
centered on the rapid rotational streaming in characean
cells. However, streaming has also been studied in dascycla-
dalean and caulerpalean algae (see following). Wound-induced contractions have been investigated primarily in si-
phonocladalean algae.

Two fundamental possibilities exist for myosin-based
cytoplasmic movement in algal and other non-muscle cells. A
myosin coating over organelles could produce unidirectional
sliding on stationary actin filaments (and vice versa?), or
bipolar myosin filaments could bring about movement of structures attached to oppositely-oriented actin filaments much
like what occurs in the muscle sarcomere (Sheetz & Spudich,
1983b). Current hypotheses suggest that each of these mechanisms may be operating separately in cytoplasmic streaming and wound-induced cytoplasmic contractions, respectively. Only an overview will be presented here, since
several reviews on streaming in the Characeae have appeared
recently (e.g., Kamiya, 1981, 1986; Shimmen & Tazawa, 1986;
Tazawa & Shimmen, 1987; Williamson, 1986).

Cytoplasmic Streaming in Charophytes

The large, multinucleate cells of Chara and Nitella
consist of a stationary cortical cytoplasm (ectoplasm), a
rapidly streaming endoplasm, and a large central vacuole.
In the cortex are slightly helical, longitudinal files of
chloroplasts, with each file containing one to a few prominent bundles of actin MFs (Fig. 2A). Within each bundle,
the MFs are parallel and in register (Palevitz & Hepler,
1975), with their "plus" ends oriented toward the downstream
direction of streaming (Kersey et al., 1976). The endoplasm
contains nuclei and other organelles that rapidly stream
basipetally in one hemicylinder, and acropetally in the
other. Transverse streaming at the ends of the cell result
in overall cyclosis throughout the cell. Between the two
hemicylinders is a clear area known as the "null zone" or

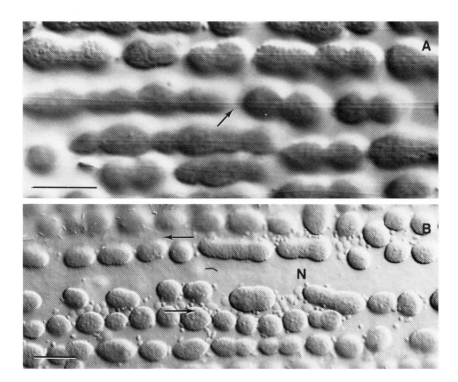

Fig. 2. Cell cortex of <u>Nitella</u> <u>axillaris</u>. A) Fine longitudinal cables of actin (arrow) attached to inner surfaces of chloroplast files represent ectoplasm-endoplasm interface. X 2050   B) Opposite directions of endoplasmic streaming (arrows) occur on either side of null zone (<u>N</u>). X 1290 Scale = 10 μm.

"neutral zone" (Fig. 2B), where the direction of streaming reverses. The current hypothesis is that streaming is caused by an active shearing force at the interface between the endoplasm and cortex, resulting from the interaction of the stationary actin bundles with endoplasmic myosin (Williamson, 1980; Kamiya, 1986). Myosin-like thick filaments associated with membrane-bound organelles have been observed with electron microscopy in <u>Chara</u> (Willamson, 1979). However, the actual location of myosin (or myosin-like proteins), including whether it is associated with any subcellular structures in these cells, is unknown. Theoretically,

myosin could be soluble, polymerized into filaments, incorporated into cytoskeletal or membranous networks or gels, or
any combination of these (Bereiter-Hahn, 1987).

There is strong evidence that particles can be transported along the actin cables, or vice versa. Isolated
actin bundles from Nitella actively slide in vitro in the
presence of Mg-ATP (Higashi-Fujime, 1980). Also, streaming
has been reconstituted on the actin bundles in tonoplast-
free and plasma-membrane permeabilized characean cells (see
Tazawa & Shimmen, 1987). Microscopic beads coated with
muscle myosin travel along actin bundles in vitro in Nitella
and Chara (Sheetz & Spudich, 1983a, b; Shimmen & Yano,
1984). In fact, these systems provide motility assays for
myosin, and experiments using these assays are greatly affecting current ideas about cytoplasmic streaming. The fact
that filamentous myosin is not necessary for movement in
vitro indicates that it is probably not essential for cytoplasmic streaming in vivo (Hynes et al., 1987; Kachar &
Reese, 1988). Furthermore, there is evidence that single-
headed myosin will support movement in vitro (Harada et al.,
1987; Toyoshima et al., 1987). Thus, the organization (non-
filamentous?) and actual nature (single-headed?) of myosin
in characean cells could be very different from such features of muscle myosin.

Most models of the mechanism of cytoplasmic streaming
in characean cells implicate unidirectional movement of
myosin (whether free in the endoplasm or attached to organelles) along the actin bundles (see Kamiya, 1981; Williamson, 1980). Ultrastructural studies indicate that in the
absence of ATP, various membranous organelles are attached
to the actin bundles as if in a state of rigor (Williamson,
1975; Nagai & Hayama, 1979). In fact, an extension of this
work has recently revived a hypothesis that a complex network of endoplasmic reticulum (ER) in the endoplasm slides
(via myosin?) along the actin bundles, providing force
enough to move the entire endoplasm, including regions well
away from the bundles (Kachar & Reese, 1988). A sliding ER
model would also satisfy earlier theoretical studies which
had indicated that an extensive fibrous or membranous network would be necessary to account for the hydrodynamic
features of streaming in the Characeae (Nothnagel & Webb,
1982).

Other models for actomyosin-based cytoplasmic streaming

in the Characeae (and in general) have been proposed. For example, Allen (1976) postulated that "undulating filaments" observed in the endoplasm of Nitella were providing the motive force for streaming. It is possible that these "filaments" represented the ER network described recently by Kachar and Reese (1988). Also a screw-type mechanical model has been proposed for streaming (Foissner, 1985), based on the model that thick and thin filaments in muscle move on each other with a screw-like motion (Jarosch & Foissner, 1982; Obendorf, 1981).

It should be noted that streaming has been studied in members of two other orders of coenocytic green algae. In Acetabularia (Dasycladales), the multistriate streaming appears to be similar to characean streaming (Koop, 1981; Fukui & Nagai, 1985). Some members of the Caulerpales have MT-based transport, including Caulerpa (Dawes & Barilotti, 1969; Kuroda & Manabe, 1983; Manabe & Kuroda, 1984), Chloro-desmis (Menzel, 1985) and Dichotomosiphon (Maekawa et al., 1986). However, Bryopsis requires both MTs and actin MFs for motility (Menzel & Schliwa, 1986). Because of the partial or exclusive role of MTs in streaming in the Caulerpales, they will not be discussed further.

Wound-induced Cytoplasmic Contractions in Siphonocladales

Similar to the charophytes, members of the Siphono-cladales also consist of a large central vacuole surrounded by a thin layer of cytoplasm. Cytoplasmic streaming is absent in siphonocladalean algae (La Claire 1984a), possibly due to the extensive arrays of microtubules reported for some of these algae (La Claire 1987, Shihira-Ishikawa 1987). However, it has been known for some time that these algae heal wounds via putative contractile mechanisms. Similar to the giant egg cells of amphibians (see Merriam & Christen-sen, 1983) and the giant amoebae (see Szubinska, 1978), puncture wounds or transverse cuts close by centripetal contractions of the cell cortex in some Valonia species (Klemm, 1894), Ernodesmis and others (La Claire, 1982a). In other species of Valonia (Doyle, 1935; Kopac, 1933; Murray, 1893; Steward & Martin, 1937) and in Boergesenia (Enomoto & Hirose, 1972), wounding induces contractions throughout the cell cortex, leading to a reticulation and subsequent sepa-ration of the cytoplasm into numerous spherical protoplasts. An intermediate pattern of motility is exhibited by Cladophoropsis, which contracts centripetally at several places along its length, producing a small number of sausage-shaped protoplasts after wounding (La Claire, 1982a). In all cases, the extensive cytoplasmic contractions lead to one or more viable multinucleate protoplasts, which may serve to propagate the organisms vegetatively. It should be added that some members of the Caulerpales also close wounds via centripetal contractions (Klemm, 1894; Burr & West, 1971; Dreher et al., 1978).

Far less is known about the mechanism of wound-induced contractions compared to streaming. Severing the base of an Ernodesmis cell induces longitudinal contractions that pull the cut end up to several millimeters from the wound site, and the cut end closes centripetally to heal the wound within 30 min (La Claire, 1982a). Most of the work in my laboratory focuses on Ernodesmis, as contractions are fast enough to be observed using a dissecting microscope, yet slow enough to manipulate experimentally. Actin is known to occur, but nothing is known yet about the presence or loca- tion of any myosin-like proteins (La Claire, 1984a). Con- tractions have been induced in permeablized cells, in which cortical bundles of MFs become evident beneath the plasma membrane solely in contracting regions of the cells, and then disappear after healing is completed (La Claire, 1984b). MF diameter is similar to that of filamentous actin, and these probably represent the actin-containing MFs extracted from contracting cells, which label with myosin (La Claire, 1984a). These bundles also specifically label with antibodies to actin and during the course of contrac- tion, they appear to shorten and thicken (La Claire, 1988). Although this shortening suggests that the MFs within the bundles are anti-parallel in Ernodesmis, the actual polarity is unknown. Since the MF bundles are temporally and spa- tially associated with contraction, and because MTs are not necessary for normal contractions (La Claire, 1987), it is hypothesized that wound-induced contractions in these algae are actomyosin-based, as shown for various animal cells (see Merriam & Christensen, 1983). Although no thick filaments have been observed yet in thin section or critical-point dried cells, some type of supramolecular myosin assembly must be involved - minimally consisting of two myosin mole- cules tail-to-tail, to bind two anti-parallel MFs. An im- portant distinction between streaming in charophytes and contractions in Siphonocladales is that in the latter the putative contractile apparatus is a transitory structure assembled, utilized and disassembled after use, whereas the actin cables function continuously in the cortex of charo-phytes.

REGULATION OF ACTIN-MYOSIN INTERACTION

Current ideas on regulatory mechanisms of actomyosin-based movement in algal cells are also based primarily on muscle models. In theory, actin-linked and/or myosin-linked regulation could be operating in any particular system. As noted above, factors that regulate the actin filament system include the availability of Ca^{2+} (and calmodulin), and the presence of troponin, tropomyosin or other actin-associated proteins. Recently, it has become clear that myosin regula-tion in many systems is related to its phosphorylation state, with either the phosphorylated or dephosphorylated form being functional, depending on the system (see Kuznicki & Barylko, 1988). In some instances, myosin itself may be directly Ca^{2+}-regulated. Although ATP is also required for motility, it generally is in abundant supply in living muscle cells, in charophytes (Williamson, 1980) and in si-phonocladalean green algae (La Claire, 1982b; 1984b), partly evidenced by the difficulty in depleting native ATP from these cell types.

Calcium Ions

Unlike most cell motility phenomena, cytoplasmic stream-ing ceases if the free Ca^{2+} concentration is equal to or greater than 1 µM (Williamson, 1975). Membrane excitation stops streaming in characean cells, due to an influx of extracellular Ca^{2+} which results in a loss of motive force generation (see Tazawa & Shimmen, 1987; Tazawa et al., 1987). In tonoplast-free cells, this Ca^{2+}-sensitive compo-nent is apparently lost or damaged, since recovered stream-ing cannot be stopped with up to 500 µM Ca^{2+}. Calmodulin (CaM) does not seem to be directly involved in regulating streaming in charophytes, but CaM inhibitors do inhibit recovery from Ca^{2+}-induced cessation in tonoplast-free cells (Tominaga et al., 1985). Experiments using foreign myosins and the coated-bead system demonstrate that beads coated with skeletal muscle myosin (which is insensitive to Ca^{2+}) move along characean actin bundles without regard to Ca^{2+} levels (see Tazawa et al., 1987; Williamson, 1986). However, if muscle tropomyosin (which does bind to the bundles) is added, Ca^{2+}-sensitivity is achieved. These experiments indicate that the native actin bundles probably do not confer Ca^{2+}-sensitivity on streaming. Similar experiments using myosin that is directly Ca^{2+}-stimulated (molluscan) or directly Ca^{2+}-inhibited (Physarum), show that characean bundles will permit movement in the presence or absence of Ca^{2+}, depending on whether the myosin itself is stimulated or inhibited (see Williamson, 1986). This work supports the concept that native characean myosin is probably the site of Ca^{2+} regulation. Although a subunit of troponin (troponin C) is the key Ca^{2+}-binding actin-regulating protein in skeletal muscle, it has not been reported yet from any plant cell. A protein immunologically similar to a different subunit (troponin T) has been localized in higher plant cells, but its distribution and function in algae are unknown (Lim et al., 1986).

Greater than 1 μM free Ca^{2+} is required to **induce** contraction in siphonocladalean cells, similar to what is found in muscle contraction and in the majority of nonmuscle motility phenomena (La Claire, 1982b, 1984b). Contraction can be mimicked and stimulated in intact Ernodesmis cells by promoting a net Ca^{2+} influx. Since La^{3+}, which competitively binds to Ca^{2+}-binding sites and Ca^{2+} channels, prevents contraction stimulation in the presence of exogenous Ca^{2+}, some type of Ca^{2+} channel might be involved with this influx (La Claire, 1983). The fact that La^{3+} also inhibits contraction in permeablized cells suggests that Ca^{2+} may directly interact with the contractile apparatus as well (La Claire, 1984a). The rapid appearance of actin bundles upon stimulation indicates that assembly of the bundles might be under Ca^{2+} control, since they do not appear in permeabilized cells inhibited by Ca^{2+}-free media (La Claire, 1984b). Recent work indicates that CaM preferentially localizes within the actin bundles, and CaM inhibitors greatly reduce contraction along with nearly eliminating actin bundle formation and shortening (Goddard & La Claire, unpublished). Collectively, these data support a role for Ca^{2+} and CaM in both the assembly and functioning of the actin bundles in Ernodesmis. However, it is not known yet whether Ca^{2+}/CaM regulation of contraction itself is actin- or myosin-linked in the Siphonocladales.

Phosphorylation

Specific enzymes that either phosphorylate or dephos-
phorylate myosin are known to play critical roles in regulating motility by (de-)activating myosin. Myosin light chain
kinases (MLCKase) and phosphatases (MLCPase) are widely
distributed and are often in turn regulated themselves by
Ca^{2+} and CaM (see Citi & Kendrick-Jones, 1987; Kuznicki &
Barylko, 1988). Phosphorylation states of myosin heavy
chains may also affect its polymerization into thick filaments. Thus phosphorylation, under the direct or indirect
control of Ca^{2+}, may regulate both the physical state and
functioning of some muscle myosins, and many non-muscle
myosins. Working with tonoplast-free characean cells, Tominaga et al. (1987) have shown that streaming is activated
even in the presence of Ca^{2+}, when some "motile component"
(possibly myosin) is dephosphorylated by a phosphatase, and
inactivated when irreversibly phosphorylated. These authors
also postulate that Ca^{2+}/ CaM might be involved in the
dephosphorylation process. Whether these events are involved in wound-induced contractions is not known.

FUTURE DIRECTIONS

To gain a deeper understanding of the mechanistics and
regulation of bulk cytoplasmic movement, it will first be
necessary to determine what (and where) other cytoskeletal
and regulatory proteins occur in these organisms (and in
plants in general). Aside from screening specific antibodies to animal proteins, the monoclonal antibody technology should prove very fruitful. First attempts at charac-
terizing monoclonal antibodies to crude cytoplasmic extracts
of Chara appear promising (Williamson et al., 1986). As
more is learned about what other proteins occur where in the
cells, biochemical characterization of these proteins will
be especially meaningful for unravelling potential regulatory mechanisms. This is particularly true for algal myosin,
since it appears not only to be the force-generating protein, but likely represents a critical component in regulating bulk movement of the cytoplasm. Determining the nature, location and regulation of plant myosin should contribute substantially to the basic understanding of intracellular movement, and giant algal cells should continue to
provide pivotal systems in this area of research.

ACKNOWLEDGEMENTS

The author is especially grateful to Doris Robinson for Fig. l, which was redrawn (with permission) from "The cooperative action of muscle proteins" ©1974 by Scientific American Inc. all rights reserved. This work was supported in part by National Science Foundation Grant DCB 84-02345 and USDA Grant 87-CRCR-1-2545.

REFERENCES

Allen NS (1976). Undulating filaments in Nitella endoplasm and motive force generation. In Goldman R, Pollard T, Rosenbaum J (eds.): "Cell Motility (Book B) Actin, Myosin and Associated Proteins", Cold Spring Harbor, NY: Cold Spring Harbor Laboratory, pp 613-621.
Allen NS, Allen RD (1978). Cytoplasmic streaming in green plants. Annu Rev Biophys Bioeng 7:497-526.
Bereiter-Hahn J (1987). Mechanical principles of architecture of eukaryotic cells. In Bereiter-Hahn J, Anderson OR, Reif WE (eds.): "Cytomechanics. The Mechanical Basis of Cell Form and Structure", Berlin: Springer-Verlag, pp 3-30.
Burr FA, West JA (1971). Protein bodies in Bryopsis hypnoides: their relationship to wound-healing and branch septum development. J Ultrastruct Res 35:476-498.
Citi S, Kendrick-Jones J (1987). Regulation of non-muscle myosin structure and function. BioEssays 7:155-159.
Dawes CJ, Barilotti DC (1969). Cytoplasmic organization and rhythmic streaming in growing blades of Caulerpa prolifera. Am J Bot 56:8-15.
Doyle WL (1935). Cytology of Valonia. Carnegie Inst Washington Papers Tortugas Lab 29:13-21.
Dreher TW, Grant BR, Wetherbee R (1978). The wound response in the siphonous alga Caulerpa simpliciuscula C. Ag.: fine structure and cytology. Protoplasma 96:189-203.
Enomoto S, Hirose H (1972). Culture studies on artificially induced aplanospores and their development in the marine alga Boergesenia forbesii (Harvey) Feldmann (Chlorophyceae, Siphonocladales). Phycologia 11:119-122.
Foissner I (1985). The rotation model as a basis for Nitella filament-dynamics. In Alia EE, Arena N, Russo MA (eds.): "Contractile Proteins in Muscle and Non-muscle Cell Systems. Biochemistry, Physiology, and Pathology", New York: Praeger Scientific, pp 213-218.

Fukui S, Nagai R (1985). Reactivation of cytoplasmic streaming in a tonoplast-permeabilized cell model of Acetabularia. Plant Cell Physiol 26:737-744.

Harada Y, Noguchi A, Kishino A, Yanagida T (1987). Sliding
movement of single actin filaments on one-headed myosin
filaments. Nature 326:805-808.

Higashi-Fujime S (1980). Active movement in vitro of bundles
of microfilaments isolated from Nitella cell. J Cell Biol
87:569-578.

Hynes TR, Block SM, White BT, Spudich JA (1987). Movement of
myosin fragments in vitro: domains involved in force production. Cell 48:953-963.

Jarosch R, Foissner I (1982). A rotation model for microtubule and filament sliding. Eur J Cell Biol 26:295-302.

Kachar B, Reese TS (1988). The mechanism of cytoplasmic
streaming in characean algal cells: sliding of endoplasmic
reticulum along actin filaments. J Cell Biol 106:1545-1552.

Kamiya N (1981). Physical and chemical basis of cytoplasmic
streaming. Annu Rev Plant Physiol 32:205-236.

Kamiya N (1986). Cytoplasmic streaming in giant algal cells:
a historical survey of experimental approaches. Bot Mag
99:441-467.

Kato T, Tonomura Y (1977). Identification of myosin in
Nitella flexilis. J Biochem (Tokyo) 82:777-782.

Kersey YM, Hepler PK, Palevitz BA, Wessells NK (1976).
Polarity of actin filaments in characean algae. Proc Natl
Acad Sci USA 73:165-167.

Klemm P (1894). Ueber die Regenerationsvorgange bei den
Siphonaceen. Bot Zeit 78:19-40.

Koop U (1981). Protoplasmic streaming in Acetabularia. Protoplasma 109:143-157.

Kopac MJ (1933). Physiological studies on Valonia ventricosa. Carnegie Inst Washington Yearb 32:273-276.

Kuroda K, Manabe E (1983). Microtubule-associated cytoplasmic streaming in Caulerpa. Proc Jpn Acad 59:131-134.

Kuznicki J, Barylko B (1988). Phosphorylation of myosin in
smooth muscle and non-muscle cells. In vitro and in vivo
effects. Int J Biochem 20:559-568.

La Claire JW II (1982a). Cytomorphological aspects of wound
healing in selected Siphonocladales (Chlorophyceae). J
Phycol 18:379-384.

La Claire JW II (1982b). Wound-healing motility in the green
alga Ernodesmis: calcium ions and metabolic energy are
required. Planta 156:466-474.

La Claire JW II (1983). Inducement of wound motility in
intact giant algal cells. Exp Cell Res 145:63-69.

La Claire JW II (1984a). Actin is present in a green alga that lacks cytoplasmic streaming. Protoplasma 120:242–244.

La Claire JW II (1984b). Cell motility during wound healing in giant algal cells: contraction in detergent-permeabilized cell models of Ernodesmis. Eur J Cell Biol 33:180–189.

La Claire JW II (1987). Microtubule cytoskeleton in intact and wounded coenocytic green algae. Planta 171:30–42.

La Claire JW II (1988). Actin cytoskeleton in intact and wounded coenocytic green algae. Planta (in press).

Lim S-S, Hering GE, Borisy, GG (1986). Widespread occurrence of anti-troponin T crossreactive components in non-muscle cells. J Cell Sci 85:1–19.

Maekawa T, Tsutsui I, Nagai R (1986). Light-regulated translocation of cytoplasm in green alga Dichotomosiphon. Plant Cell Physiol 27:837–851.

Manabe E, Kuroda K (1984). Ultrastructural basis of microtubule associated cytoplasmic streaming in Caulerpa. Proc Jpn Acad 60:118–121.

Menzel D (1985). Fine structure study on the association of the caulerpalean plastid with microtubule bundles in the siphonalean green alga Chlorodesmis fastigiata (Ducker, Udoteaceae). Protoplasma 125:103–110.

Menzel D, Schliwa M (1986). Motility in the siphonous green alga Bryopsis. II. Chloroplast movement requires organized arrays of both microtubules and actin filaments. Eur J Cell Biol 40:286–295.

Merriam RW, Christensen K (1983). A contractile ring-like mechanism in wound healing and soluble factors affecting structural stability in the cortex of Xenopus eggs and oocytes. J Embryol Exp Morphol 75:11–20.

Murray G (1893). On Halicystis and Valonia. In Murray G (ed.): "Phycological Memoirs. Part II", London: Dulau, pp 47–52.

Murray JM, Weber, A (1974). The cooperative action of muscle proteins. Sci Am 230(2):58–71.

Nagai R, Hayama T (1979). Ultrastructure of the endoplasmic factor responsible for cytoplasmic streaming in Chara internodal cells. J Cell Sci 36:121–136.

Nothnagel EA, Webb WW (1982). Hydrodynamic models of viscous coupling between motile myosin and endoplasm in characean algae. J Cell Biol 94:444–454.

Obendorf, P (1981). A rotating myosin filament theory of muscular contraction. J Theor Biol 93:667–680.

Palevitz BA, Hepler PK (1975). Identification of actin in situ at the ectoplasm–endoplasm interface of Nitella. Microfilament-chloroplast association. J Cell Biol 65:29–38.

Sheetz MP, Spudich JA (1983a). Movement of myosin-coated fluorescent beads on actin cables in vitro. Nature 303:31–35.

Sheetz MP, Spudich JA (1983b). Movement of myosin-coated structures on actin cables. Cell Motil 3:485–489.

Shihira-Ishikawa I (1987). Cytoskeleton in cell morphogenesis of the coenocytic green alga Valonia ventricosa. I. Two microtubule systems and their roles in positioning of chloroplasts and nuclei. Jpn J Phycol 35:251–258.

Shimmen T, Tazawa M (1986). Control mechanism and reconstitution of cytoplasmic streaming in Characeae. In Ishikawa H, Hatano S, Sato H (eds.): "Cell Motility: Mechanism and Regulation", New York: Alan R. Liss, pp 253–261.

Shimmen T, Yano M (1984). Active sliding movement of latex beads coated with skeletal muscle myosin on Chara actin bundles. Protoplasma 121:132–137.

Steward FC, Martin JC (1937). The distribution and physiology of Valonia at the Dry Tortugas, with special reference to the problem of salt accumulation in plants. Carnegie Inst Washington Papers Tortugas Lab 31:87–170.

Szubinska B (1978). Closure of the plasma membrane around microneedle in Amoeba proteus: an ultrastructural study. Exp Cell Res 11:105–115.

Taylor EW (1986). Cell motility. J Cell Sci Suppl 4:89–102.

Tazawa M, Shimmen T (1987). Cell motility and ionic relations in characean cells as revealed by internal perfusion and cell models. Int Rev Cytol 109:259–312.

Tazawa M, Shimmen T, Mimura T (1987). Membrane control in the Characeae. Annu Rev Plant Physiol 38:95–117.

Tominaga Y, Muto S, Shimmen T, Tazawa M (1985). Calmodulin and Ca^{2+}-controlled cytoplasmic streaming in Characean cells. Cell Struct Funct 10:315–325.

Tominaga Y, Wayne R, Tung HYL, Tazawa M (1987) Phosphorylation-dephosphorylation is involved in Ca^{2+}-controlled cytoplasmic streaming of characean cells. Protoplasma 136:161–169.

Toyoshima YY, Kron SJ, McNally EM, Niebling KR, Toyoshima C, Spudich JA (1987). Myosin subfragment-1 is sufficient to move actin filaments in vitro. Nature 328:536–539.

Vorob'eva IA, Pogalzov BF (1963). Isolation of contractile protein from the alga Nitella flexilis. Biofizika 8:427–429.

Williamson RE (1975). Cytoplasmic streaming in Chara: a cell model activated by ATP and inhibited by cytochalasin B. J Cell Sci 17:655–668.

Williamson RE (1979). Filaments associated with the endoplasmic reticulum in the streaming cytoplasm of Chara corallina. Eur J Cell Biol 20:177–183.

Williamson RE (1980). Actin in motile and other processes in plant cells. Can J Bot 58:766-772.

Williamson RE (1986). Organelle movements along actin filaments and microtubules. Plant Physiol 82:631-634.

Williamson RE, Perkin JL, McCurdy DW, Craig S, Hurley U (1986). Production and use of monoclonal antibodies to study the cytoskeleton and other components of the cortical cytoplasm of Chara. Eur J Cell Biol 41:1-8.

Williamson RE, McCurdy DW, Hurley UA, Perkin JL (1987). Actin of Chara giant internodal cells. A single isoform in the subcortical filament bundles and a larger, immunologically related protein in the chloroplasts. Plant Physiol 85:268-272.

Algae as Experimental Systems pages 71–91
© **1989 Alan R. Liss, Inc.**

MAINTENANCE AND DYNAMIC CHANGES OF CYTOPLASMIC ORGANIZATION
CONTROLLED BY CYTOSKELETAL ASSEMBLIES IN ACETABULARIA
(CHLOROPHYCEAE)

Diedrik Menzel and Christine Elsner-Menzel

Max-Planck-Institut für Zellbiologie,
Rosenhof, D-6802 Ladenburg, F.R.G.

The advantages of the giant, unicellular green alga
Acetabularia, as an experimental system in cell biological
research, has been widely documented and summarized in many
recent review articles (Bonotto et al., 1976; Berger et al.,
1987; Neuhaus & Schweiger, 1987). This article specifically
focuses on cytoplasmic morphogenesis, including mechanisms
utilized by the cell to actively maintain and change the
shape of designated cytoplasmic domains in the course of
development.

The phenomenon of cytoplasmic shaping, which occurs
during formation of cysts at the end of the cell's life
cycle, has been known for more than 50 years (Schulze,
1939). The fundamental principle underlying this seemingly
simple process was recognized by Werz (1968, 1969b, 1970),
who demonstrated that the shape of the cell is not determined by the cell wall but rather by the protoplast prior to
cell wall formation. The cell wall merely supports and en-
hances the final shape. This principle is now widely accep-
ted for plant cells (e.g., Kiermayer, 1981; Lloyd, 1982).

The observation that the alkaloid colchicine blocks
specific stages of morphogenesis led Werz (1970) to hypothe-
size the involvement of cytoskeletal elements in the regulation of morphogenesis long before this concept emerged from
studies on higher plants (Heath & Seagull, 1982). Recent
fine structural and immunochemical data confirm this role of
the cytoskeleton in morphogenesis of Acetabularia.

THE VEGETATIVE PHASE

During much of the alga's life cycle (Fig. 1; also see
Puiseux-Dao, 1970) two morphogenetic programs alternate at
the growing tip: polar elongation along the main axis, and
initiation of side branch primordia. A switch from one program to the next results in a 90 degree shift in the growth
axis. Each primordium differentiates into several ranks of
trifurcate branches, making up a whorl. Alternation of tip
growth with side branch formation is repeated several times
before a third morphogenetic program takes over, which is
also accompanied by a 90 degree rotation of the primary
growth axis. This time however, much larger side branch primordia form and differentiate into the gametangial rays.

Goodwin & Trainor (1985) postulate the involvement of a
"cytogel template" in the control of the position of primordia. The geometry of this template is assumed to be determined by radially symmetrical modulations in cytoplasmic
Ca^{2+} at the subapical initiation sites. In confined regions
of high Ca^{2+} concentration, interactions among putative

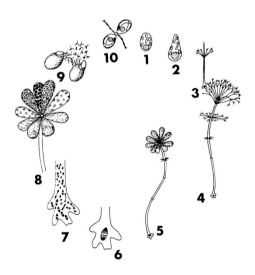

Fig. 1) Life cycle of <u>Acetabularia</u> <u>(Polyphysa)</u> <u>cliftoni</u>. See
text for details. Not drawn to scale.

cytoskeletal elements are assumed to weaken the viscoelastic properties of the cytogel. Support for the influence of Ca^{2+} comes from two observations. 1) the number of primordia decreases with decreasing external Ca^{2+} concentrations; 2) preliminary observations reveal Ca^{2+} localized in a sub-apical annular ring prior to whorl formation (Harrison & Hillier, 1985). Although the regulatory function of Ca^{2+} in cytoskeletal interactions is well established in other organisms (Hepler & Wayne, 1985; Trewavas, 1986; Williamson, 1984), the operation of such a mechanism in Acetabularia remains speculative.

It is remarkable that colchicine inhibits only the relatively brief morphogenetic transition, i.e. the change from tip elongation to side branch initiation; whereas growth before and after this event is not affected at all (Werz, 1969a & b). However, very little is known about spatial cytoplasmic organization in the growth zones. As yet there is no supporting evidence that microtubules (MTs) control the shift of the growth axis. MTs are not detectable in the stalk cytoplasm by immunofluorescence cytochemistry (Menzel, 1986).

The cytoplasm in the vegetative phase is dominated by a cytoskeleton composed of massive bundles of actin microfilaments embedded in a fine network of actin filaments. The bundles form a parallel, axially oriented system running along the entire length of the cell, including the cap rays (Dazy et al., 1981; Menzel, 1986). Chloroplasts and other organelles are transported along these bundles (filament type transport). Koop & Kiermayer (1980a) described an additional transport mode for organelles, which they termed "headed streaming bands" (HSB). It is analogous to the fast lane of traffic on an expressway; polyphosphate granules and a host of smaller ill-defined vesicles move along at a rate of up to three times the velocity of ordinary chloroplast transport. The mechanisms underlying the two different modes of transport are not identified, although many experiments provide circumstantial evidence that the cytoskeleton is involved. In the case of chloroplast transport, the evidence points to an actin/myosin based mechanism (Koop & Kiermayer, 1980b; Nagai & Fukui, 1981). The motor which drives the HSB, remains mysterious. Unlike chloroplast transport, HSB-movement port is reversibly inhibited by colchicine and the MT specific herbicide amiprophosmethyl. The observation seemingly argues for an involvement of MTs, but immunofluores-

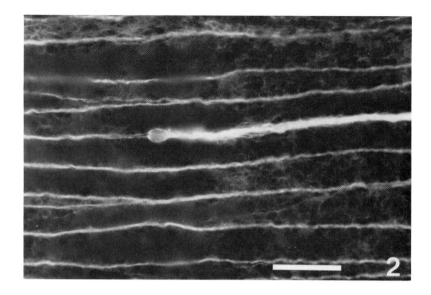

Fig. 2) Part of stalk cytoplasm prior to meiosis: actin
bundles visualized by indirect immunofluorescence using a
monoclonal antibody against smooth muscle actin. HSB in
center; direction of movement was to left. Scale 50 μm.

cent data do not support this. However, if MTs are present
in minute quantities, or too ephemeral to be visualized by
present methods, it is still possible that these apparently
opposing data can be reconciled. It may also be that the
antigen (tubulin) is masked within the HSB by tightly bound
macromolecules and therefore unrecognizable to the antibodies. Hence, the issue of MT presence in HSBs remains
open.

    Cinematographic and time lapse video studies of head
movements reveal that the HSB-head has a mechano-elastic
behavior much like an amoeboid pseudopodium (Koop, 1979;
Menzel, unpublished). The cytoskeletal organization of the
head reveals a diffuse distribution of actin, possibly a
fine meshwork unresolvable by light microscopy (Fig. 2). The
head is perfectly superimposed on the largest actin bundles
within the bundle system. Typically the diameter of the

actin bundle that carries a head structure, is smaller in
the moving direction of the head than behind it (Fig. 2).
This suggests that massive actin filament polymerization
accompanies the moving head.

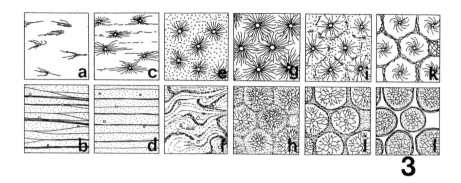

Fig. 3) Diagram of cytoskeletal changes during cyst morphogenesis (details in text); distribution of MTs is shown in
the upper row, distribution of actin microfilaments in the
lower row; a,b) nuclear migration in the stalk; c,d) intra-
cap ray migration; e,f) mixing phase; g,h) disk stage; i,j)
ring stage; k,l) dome stage; original data from Menzel
(1986, 1988).

TRANSLOCATION OF NUCLEI

     The vegetative phase ends with the division of the pri-
mary nucleus probably undergoing meiosis (Koop et al.,
1979). Spindle formation during meiosis marks the first
verifiable appearance of MTs in the life cycle of Acetabularia (Fig. 1); it is followed by the formation of several
thousand secondary nuclei, each of them furnished with a MT
system, which usually radiates asymmetrically from one side
of the nucleus into the surrounding cytoplasm (much like a
comet's tail, Fig. 3a). In this phase of the life cycle,
overall organelle movement is directed towards the cap
(Schulze, 1939). Nuclei move somewhat faster along the HSB-
lane than other organelles (Koop & Kiermayer, 1979) and
eventually collect within the HSB-head (Menzel, unpublished). The reason for this peculiar motile behavior is not
known. Perinuclear MTs may be involved in both fast trans-

port as well as entrapment of nuclei within the head. Although a hybrid molecular motor between actin filaments and MTs mediating directional force production has not been discovered, interaction between the two cytoskeletal elements is becoming an attractive model that is being discussed for both animal and plant cells (Kobayashi et al., 1988; Menzel & Schliwa, 1986b; Uyeda & Furuya, 1987; Vasiliev, 1987). A likely candidate for a molecular linker between actin and MTs is MAP-2, one of several high molecular weight MT associated proteins (Pollard et al., 1984). However, it is not known if this protein is involved in force production. Acetabularia could provide a source for the isolation of such a hybrid motor protein.

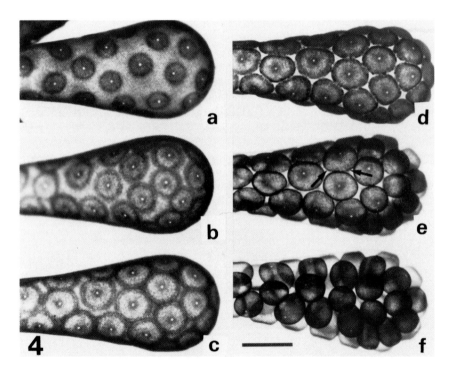

Fig. 4) Stages of cyst morphogenesis (A. cliftoni); scale 500 μm. a) Disk stage; b) early ring stage; c) late ring stage; d) early dome stage; e) late dome stage, arrows show position of ring; f) end of contraction stage.

CYST MORPHOGENESIS

As secondary nuclei enter the cap rays, dramatic changes occur in cytoplasmic architecture prior to cyst formation (Schulze, 1939; Werz, 1970; some of these stages are
shown in Fig. 4). In the first stage secondary nuclei
migrate into the cytoplasm of the cap rays in more or less
regular rows. Chloroplast movement is nearly normal and the
actin cytoskeleton looks similar to the stalk (Fig. 3d).
However, the perinuclear MT rays are much more extended and
symmetrically arranged than in the stalk with the longest MT
marking the long axis of the cap ray (Fig. 3c, 5); they also
frequently align along the most prominent actin bundles
(Fig. 3c, d). This suggests that the migration of nuclei
within the cap rays is the result of interactions between
the perinuclear MTs and the actin bundles as suggested for
nuclear movement in the stalk. Differences in motile behavior may be due to the larger MT systems in the cap rays.

As cap rays enter the "mixing phase" the normal pattern
of chloroplast movement is replaced by a more erratic mode
of movement along a few constantly changing and slanted
streamlets (Koop, 1978; Schulze, 1939). Direction of movement
frequently reverses and contraction waves may progress
through the cytoplasm. The velocity of chloroplast movement
is drastically reduced. Most chloroplasts become immobile
and densely packed. Nuclei also change motile behavior.
Instead of proceeding along defined tracks, adjacent nuclei
start to randomly swap positions. Time lapse video recor-
dings reveal that forces from varying directions seem to
transiently pull at the nuclei (Koop, 1978; Menzel, unpublished).

Changes in organelle movement and distribution are
reflected by changes in cytoskeletal organization. Arrays of
straight and parallel actin filament bundles disappear,
giving way to a great number of curved, elongate actin bands
embedded within a fine meshwork of actin filaments (Fig. 6).
The perinuclear MT systems further increase and attain a
radially symmetrical alignment. Spaces between the nuclei
are dotted with small uniform MT fragments (Fig. 3e, 7).
Apparently, polymerization of MTs occurs in several phases.
The first is the formation of perinuclear MTs shortly after
production of secondary nuclei. The second phase is represented by the extension of the perinuclear MT systems during
intra-cap ray migration of nuclei (Fig. 3a, c). In both, MTs

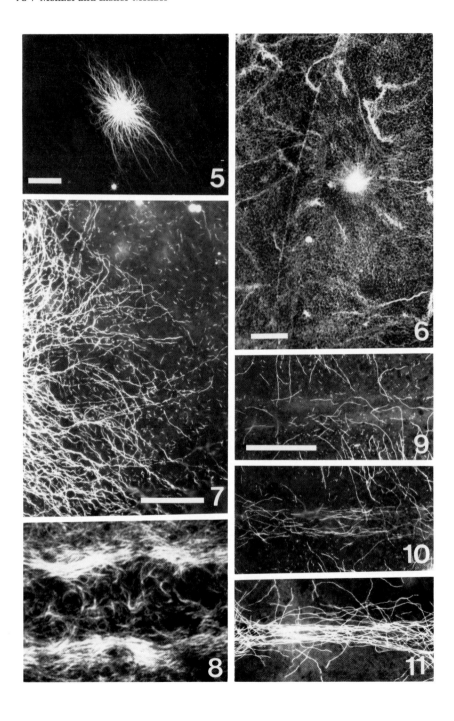

remain closely associated with the nuclear envelope and assembly appears to be controlled from an area at or close to the nuclear envelope. During the mixing phase the third wave of MT assembly occurs by random polymerization in the cytoplasm (Fig. 7). The reason for this difference is unclear. Fragments may have sloughed off from perinuclear MTs during the mixing process or, alternatively, the tubulin concentration could have transiently risen dramatically in the cytoplasm to match nucleation conditions. Preliminary evidence does not favor a random MT nucleation capacity in the cytoplasm in the absence of MT seeds. This conclusion has been drawn from studies on the reappearance of MTs in the presence or absence of 10 μM CIPC (isopropyl N-[3-chlorophenyl] carbamate) after they had been completely broken down by the MT specific inhibitor oryzalin. The carbamate derivative, which is known to interfere with MTOC activity (Clayton & Lloyd, 1984; Fedtke, 1982), prevented MT reappearance, whereas in the controls MTs reformed exclusively around the nuclei. This could be taken as an indication for the absence of MTOC-like activity elsewhere in the cytoplasm except for the nuclear membrane and the perinuclear cytoplasm (Menzel, unpublished).

As the process of mixing gradually subsides, chloroplasts gather around the nuclei in circular domains typical of the disk stage (Fig. 4a). The perinuclear MT systems have reached their largest expansion (Fig. 3g). Since MT fragments have disappeared it may be legitimate to assume that they have been incorporated into the radial MT systems by annealing, a phenomenon that is well known for MTs in vitro (Rothwell et al., 1986). The disk stage also marks the beginning of actin ring assembly by a condensation of the fine network of actin filaments around the circumference of

---

Fig. 5) Perinuclear MT system of secondary nucleus migrating in cap ray cytoplasm, indirect immunofluorescence using a monoclonal anti-chicken tubulin; scale 100 μm. Fig. 6) irregular actin bands around a secondary nucleus (bright spot in center); mixing stage; scale 100 μm. Fig. 7) MT fragments at margin of perinuclear MT system ; mixing stage; domain center to left; scale 50 μM. Fig. 8) Small curving actin bundles in gap between forming rings; scale 10 μm. Fig. 9-11) Formation of circumferential MT system in area of ring; neighboring cyst domains are to top and bottom; scale 50 μm. 9) MT fragments begin to align parallel to ring. 10) Increase of MTs in ring area. 11) Completion of MT assembly at dome stage.

the disks (Fig. 3h).

The transit from disk to ring stage is macroscopically characterized by a partial relocation of chloroplasts from the disk centers to the periphery, resulting in the transient appearance of green rings around the nuclei (Fig. 4b). This event coincides with the formation and gradual alignment of short actin bundles around the circumference of the disks (Fig. 3j, 8). At the same time the perinuclear MT systems undergo another partial breakdown at the distal ends producing MT fragments, which become incorporated into a second MT system that co-localizes with the forming actin rings (Fig. 9, 10, 11). Thus the domains of the future cysts become delineated by two interlaced cytoskeletal systems.

Up until the time of actin ring assembly, the cytoplasm remains smoothly appressed to the cell wall. However, as ring assembly proceeds to completion the cyst domains (defined as the designated area of cap ray cytoplasm that will be transformed into a cyst) gradually bulge outward and the plane of the rings sinks inward. By this process each domain is slowly transformed into a dome (dome stage; Fig. 4d). It is only at this stage that chloroplasts, which have aligned along the rings, simultaneously start to move very slowly (a few micrometers per minute) in a clockwise direction for several tens of minutes (Menzel, unpublished). The homogeneous direction of movement points at a homogeneous polarity of the underlying actin fiber bundles provided that the mechanism of chloroplast movement is based on actin cables and organelle-bound myosin (Lackie, 1986; Sheetz et al., 1986).

At the contraction stage cytoskeletal organization appears to be consolidated. The network begins to function as an integrated mechanism designed for the final shaping process involving a reduction in ring diameter and an increase in surface area. During "contraction" of the actin rings the perinuclear MT systems further recede and MTs become somewhat distorted with the distal MT ends curving to the left (Fig. 3k, 18). Since this pattern is consistently present in all domains during this stage it is likely to be functionally linked to the contraction process. The increase in surface area over each domain is probably mediated by large numbers of coated vesicles observed in stages of apparent fusion with the plasmalemma and the tonoplast (Franke et al., 1977). By the time ring contraction is nearing completion a stem-like structure has formed at the base of each cyst protoplast. These stems are the contracted remnants of the rings which eventually rupture leaving a contracted mass of cytoplasm behind in the center region of the cap ray with completed cyst protoplasts lined up peripherally beneath the cell wall.

PHARMACOLOGICAL PERTURBATION

Inhibitor studies combined with electron microscopy have suggested an involvement of MTs in the positioning of nuclei (Werz, 1969a; Woodcock, 1971), and of actin in the contraction phase respectively (Zimmer & Werz, 1981). Recent advances in the visualization of cytoskeletal elements allow a more meaningful analysis of pharmacological experiments, because now the effects of drugs can be more easily correlated with changes in the organization of the cytoskeleton (Menzel & Schliwa, 1986a).

A series of inhibitor studies using the MT-specific herbicide oryzalin, the fungal metabolite cytochalasin D (CD), the SH-blocker N-ethylmaleimide (NEM) and the adenosine analog erythro-9-[3-(2-hydroxynonyl)]adenine (EHNA) has lead to fundamental insights into cytoskeletal dynamics and has unraveled some of the structural complexity of the cytoskeletal assemblies involved. These studies suggest that MTs not only control nuclear positioning but also play a crucial role in defining the domain borders and spatially restrict actin bundle assembly to the future ring area. They further suggest that MTs are needed to maintain tension over the domain area, once the actin rings are completed.

By treatment with CD the actin rings (Fig. 12) can be removed selectively, whereas the actin filament network over the cyst domains persists and even appears strengthened in the presence of the inhibitor (Fig. 13). But the network alone does not suffice to produce tension. Cyst domains devoid of rings cannot proceed into the dome stage. If CD is applied when domes have already formed, they flatten again and cyst morphogenesis stops. If both, actin rings and actin filament network are intact but MTs are absent, again, cyst morphogenesis is blocked or highly aberrant (Menzel 1988). This demonstrates, that neither the MT systems alone, nor the actin cytoskeleton alone can perform their roles in the bulging process in the absence of their respective cytoskeletal counterparts.

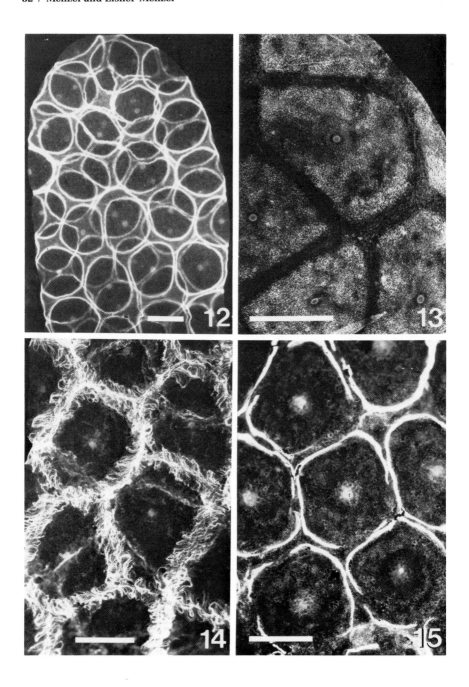

A most drastic effect in perturbing the delicate balance between MT systems and actin containing parts of the cytoskeletal network is caused by EHNA (Fig. 20c). It has been suggested previously that the drug exerts its effect by disabling a factor in the cytoplasm which controls binding between actin filament bundles (Menzel, 1988). In early stages of assembly actin rings become aberrant and remain incomplete and non functional. At the dome stage the actin rings separate into segments, which splay apart from each other (Fig. 20c). This causes a change in the spacing between adjacent ring elements and the result is that free ends of the bundle segments originating from neighboring rings start to laterally slide past one another. However, the presence of MTs prevents these ring segments from moving away from the rings. This causes a most conspicuous uniform rotation of all cyst domains in a counter-clockwise direction (Fig. 20c). The direction of movement is opposite to chloroplast movement observed around the rings in untreated cells of the same stage. Both, clockwise chloroplast movement in untreated cells and counter-clockwise domain rotation based on lateral sliding of adjacent ring elements in EHNA treated cells are consistent with the assumption that the polarity of actin filaments is uniform in all rings (minus ends facing counter-clockwise, Fig. 20b). It can be shown by treatment with EHNA plus oryzalin during the dome stage that basically the same process happens in the absence of MTs only now the actin bundle segments are free to move and actually slide past one another for some distance. This causes the domain structure to quickly deteriorate without uniform rotation (Fig. 20e).

Fig. 12) Survey micrograph of an entire cap ray showing geometry of actin rings in untreated alga at dome stage; scale 250 μm. Fig. 13) Disappearance of actin rings around cyst domains after treatment with CD (5μg/ml, 30 min); however, staining within fine, reticulate network of actin fibers over face of domains is stronger than in control; scale 250 μm. Fig. 14) Appearance of actin rings after treatment with NEM (40μM, 30 min); rings disintegrate by forming multiple loop structures; scale 250 μm. Fig. 15) Appearance of actin rings after treatment with NEM (5μM, 10 min) followed by EHNA (200μM, 1h); note discontinuities within rings; see Fig. 20 for summary of inhibitor effects. Scale 250 μm.

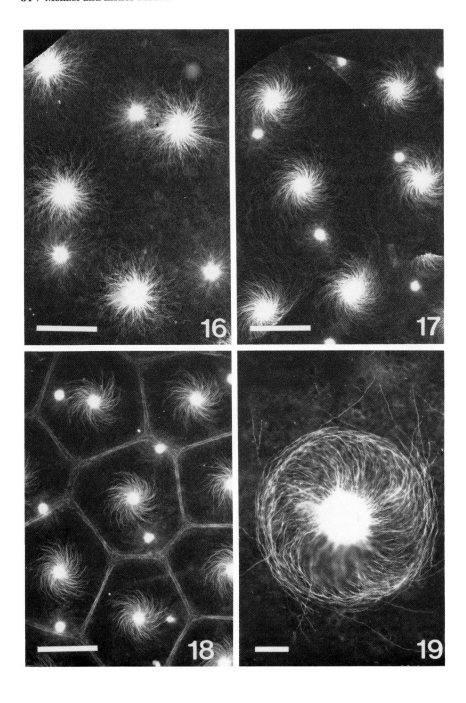

In the presence of low concentrations of NEM (10 μM), EHNA still causes fragmentation of the rings into segments (Fig. 20f) but bundle segments are not displaced from their location. This supports the hypothesis that the primary action of EHNA is a loosening of bonds between ring segments. The effects of NEM on living cells are not easy to interpret since NEM is a general SH-group blocking agent (Wallace et al., 1987) and can be expected to interfere not only with SH-enzymes like myosin (Pemrick & Weber, 1976), but also with other SH-proteins like actin and MTs (Faulstich et al., 1984; Ikeda & Steiner, 1978). It is therefore not surprising to see a partial depolymerization of MTs in the presence of NEM. Judging by the appearance of ring segments under the influence of NEM/EHNA the polymer state of actin also appears to be affected (Fig. 15). Yet the fact that these segments do not move points in favor of an impairment of the molecular motor which drives ring rotation in the presence of EHNA alone. NEM at higher concentrations, i.e. 40 μM severely disrupts actin bundle alignment within the ring (Fig. 14, 20h); probably due to a combined action on the stability of actin filaments and MTs and on the force producing myosin.

Another example illustrating the interdependence of MTs and actin filaments in building functional cytoskeletal assemblies is the premature formation of left hand MT whirls under conditions where the actin cytoskeleton is greatly disrupted (Fig. 16, 17). At the disk stage, the ends of MTs in the radially symmetrical perinuclear arrays curve in variable directions (Fig. 16). However, if cells enter the dome stage the MT ends are somehow caused to, slightly but uniformly, bend to the left side (Fig. 18). This effect can be induced earlier at the disk stage by treatment with CD (Fig. 17) and can be greatly enhanced if CD is adminis-

Fig. 16-18) Survey micrographs of adjacent secondary nuclei with their radially symmetrical MT systems. Scale 250 μm. 16) Untreated cell, disk stage; distal ends of MTs show no preferential alignment. 17) Disk stage after treatment with CD; uniform, counter-clockwise curve of distal MT ends. 18) Untreated cell at beginning of contraction phase; note uniform counter-clockwise curving of distal MT ends. Fig. 19. Perinuclear MT system at late dome stage after treatment with CD; counter-clockwise distortion of distal MT ends is enhanced; MTs almost form second ring within larger ring (not seen), scale 50 μm.

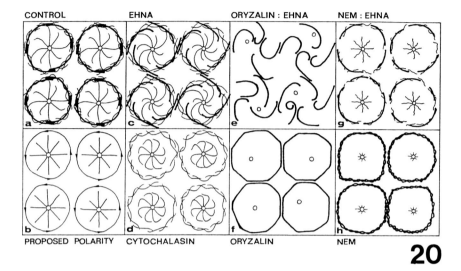

Fig. 20) Diagram summarizing cytoskeletal perturbations
after various treatments of dome stage cap rays; MTs:thin
lines, actin bundles:thick lines; a) Control: each actin
ring assumed to be composed of 4 or more segments held
together at ends. Two MT systems; (i) perinuclear system,
MTs slightly bent at distal ends, (ii) circumferential system interlaced with actin rings. b) Assumed polarity of
actin filaments in rings, arrow heads mark pointed ends;
note opposite polarity of adjacent bundles. c) Effects of
EHNA: separation of actin ring segments and lateral association of free ends resulting in simultaneous torsion of
domains. d) Effects of CD treatment: actin rings depolymerized; perinuclear MT arrays sharply bending counter-
clockwise. e) Effects of EHNA in combination with oryzalin:
MTs depolymerized, actin rings split into segments, free
ends of segments interact by lateral sliding. f) Effects of
oryzalin alone: MTs depolymerized, actin rings dilated and
compressed against each other, nuclei displaced from center
of domains. g) Pretreatment with NEM (10 µM/10 min) prior to
EHNA action: actin rings split into segments, free ends of
segments do not interact; MT arrays partly disrupted h)
Effect of NEM alone at 40 µM (30 min); rings splayed, but
segments do not split, MTs greatly depolymerized.

tered during the dome stage (Fig. 18, 19). The nature of thé factor that causes homogeneous, left-hand curving of MT ends remains completely obscure. It could be possible that the unique deformation of the original geometrical order is associated with a rotational movement of the entire arrays, however attempts to detect rotations by time lapse video microscopy have not been successful.

CONCLUSIONS

Acetabularia has proven to be an elegant model organism to study the involvement of the cytoskeleton in cytoplasmic shaping as part of cellular morphogenesis in plant cells, an area of research which is only beginning to emerge. The present data lead to the conclusion that the cell utilizes a cytoskeletal mechanism consisting of four functional elements in order to convert a flat sheath of cytoplasm into domes and ultimately into spheres: (i) a radially symmetrical MT system providing rigidity; (ii) a fine network of actin filaments over the domain face reinforcing the radial alignment of the MTs; (iii) a ring of actin filament bundles producing contractile force; and (iv) a circumferential MT-system acting as an "anchor" upon which force can be exerted. The concerted action of the four elements controls the shape change, and it is obvious that this process will become aberrant if one of the elements is experimentally tipped out of balance. The limits of the inherent geometrical order can be expected to be laid down at the beginning of the disk stage, when nuclei associate with the plasmamembrane. At that stage MTs originate from an area of cytoplasm between the nuclear envelope and the plasmamembrane and this ultimately controls the direction of bulging. Later, at the beginning of the ring stage, the perinuclear MT-systems may help define the margins of each cyst domain and thus predetermine the positioning of the future actin rings. It may therefore be generalized that cytoskeletal complexity increases stepwise during cyst formation, each step setting the stage for the next, up to the moment when a point of final differentiation is reached culminating in ring contraction.

In order to experimentally dissect functional components of the underlying mechanisms in more detail, it will be necessary to study MT and actin polarity as a component of the regular geometry of the cytoskeletal arrays as well as their potential interactions with the plasmalemma; the latter has remained completely unexplored in current research on Acetabularia. Finally, attention will have to focus on the nature of the macromolecular components which can be expected to specifically bind to the structural polymers and mediate interactions between them.

ACKNOWLEDGEMENTS

We thank Drs. J. Olsen and J. Willingale for critically reading the manuscript and for many helpful suggestions. Financial support by the Deutsche Forschungsgemeinschaft is greatly appreciated.

REFERENCES

Berger S, de Groot EJ, Neuhaus G, Schweiger M (1987). Acetabularia: A giant single cell organism with valuable advantages for cell biology. Eur J Cell Biol 44:349-370.

Bonotto S, Lurquin P, Mazza A (1976). Recent advances in research on the marine alga Acetabularia. Adv Mar Biol 14:123-250.

Clayton L, Lloyd CW (1984). The relationship between the division plane and spindle geometry in Allium cells treated with CIPC and griseofulvin: an antitubulin study. Eur J Cell Biol 34:248-253.

Dazy A-C, Hoursiangou-Neubrun D, Sauron ME (1981). Evidence for actin in the marine alga Acetabularia mediterranea. Biol Cell 41:235-238.

Faulstich H, Merkler I, Blackholm H, Stournaras C (1984). Nucleotide in monomeric actin regulates the reactivity of the thiol groups. Biochemistry 23:1608-1612.

Fedtke C (1982). Microtubules. In: Fedtke C: "Biochemistry and Physiology of Herbicide Action". Berlin: Springer Verlag, pp 123-141.

Franke WW, Spring H, Kartenbeck J, Falk H (1977). Cyst formation in some dasycladacean green algae. I. Vesicle formations during coenocytotomy in Acetabularia mediterranea. Eur J Cell Biol 14:229-259.

Goodwin BC, Trainor LEH (1985). Tip and whorl morphogenesis in Acetabularia by calcium-regulated strain fields. J Theor Biol 117:79-106.

Harrison LG, Hillier NA (1985). Quantitative control of Acetabularia morphogenesis by extracellular calcium: A test of kinetic theories. J Theor Biol 114:177-192.

Heath IB, Seagull RW (1982) Oriented cellulose fibrils and the cytoskeleton: a critical comparison of models. In Lloyd CW (ed): "The Cytoskeleton in Plant Growth and Development", London: Academic Press, pp 163-182.

Hepler PK, Wayne RO (1985). Calcium and plant development. Annu Rev Plant Physiol 36:397-439.

Ikeda Y, Steiner M (1978). Sulfhydryls of platelet tubulin: Their role in polymerization and colchicine binding. Biochem 17:3454-3459.

Kiermayer O (ed) (1981). "Cytomorphogenesis in Plants". Cell Biology Monographs, Vol. 8. Wien: Springer Verlag.

Kobayashi H, Fukuda H, Shibaoka H (1988). Interrelation between the spatial disposition of actin filaments and microtubules during the differentiation of tracheary elements in cultured Zinnia cells. Protoplasma 143:29-37.

Koop HU (1978). Entwicklung von Acetabularia (Dasycladales). Inst Wiss Film, Göttingen, Film C1298.

Koop HU, Kiermayer O (1979). Stadienspezifische Protoplasmaströmung im Stiel von Acetabularia mediterranea. Inst Wiss Film, Göttingen, Film C1384.

Koop HU, Kiermayer O (1980a). Protoplasmic streaming in the giant unicellular green alga Acetabularia mediterranea. I. Formation of intracellular transport systems in the course of cell differentiation. Protoplasma 102:147-166.

Koop HU, Kiermayer O (1980b). Protoplasmic streaming in the giant unicellular green alga Acetabularia mediterranea: II. Differential sensitivity of movement systems to substances acting on microfilaments and microtubules. Protoplasma 102: 295-306.

Koop H-U, Schmid R, Heunert H-H, Spring H (1979). Spindle formation and division of the giant primary nucleus of Acetabularia (Chlorophyta, Dasycladales). Differentiation 14:135-146.

Lackie JM (1986). "Cell Movement and Cell Behaviour". London: Allen & Unwin, chapter 2, 3.

Lloyd CW (1982). "The Cytoskeleton in Plant Growth and Development". London: Academic Press.

Menzel D (1986). Visualization of cytoskeletal changes through the life cycle in Acetabularia. Protoplasma 134:30-42.

Menzel D (1988). Perturbation of cytoskeletal assemblies in cyst domain morphogenesis in the green alga Acetabularia. Eur J Cell Biol 46:217-226.

Menzel D., Schliwa M (1986a) Motility in the siphonous green alga Bryopsis.I. Spatial organization of the cytoskeleton and organelle movements. Eur J Cell Biol 40:275-285.

Menzel D, Schliwa M (1986b). Motility in the siphonous green alga Bryopsis. II. Chloroplast movement requires organized arrays of both microtubules and actin filaments. Eur J Cell Biol 40:286-295.

Nagai R, Fukui S (1981). Differential treatment of Acetabularia with cytochalasin B and N-ethylmaleimide with special reference to their effects on cytoplasmic streaming. Protoplasma 109:79-89.

Neuhaus G, Schweiger H-G (1987). Two way traffic between nucleus and cytoplasm: cell surgery studies of Acetabularia. In Peters R, Trendelenburg M (eds): "Nucleocytoplasmic Transport". Berlin: Springer Verlag, pp. 63-71.

Pemrick S, Weber A (1976). Mechanism of inhibition of relaxation by N-ethylmaleimide treatment of myosin. Biochemistry 15: 5193-5198.

Pollard TD, Seldon SC, Maupin P (1984). Interaction of actin filaments with microtubules. J Cell Biol 99:33s-37s 2106.

Puiseux-Dao S (1970). "Acetabularia and Cell Biology". London: Logos Press.

Rothwell SW, Grasser WA, Murphy DB (1986). End-to-end annealing of microtubules in vitro. J Cell Biol 102: 619-627

Schulze KL (1939). Cytologische Untersuchungen an Acetabularia mediterranea und Acetabularia wettsteinii. Arch Protistenkunde 92:179-225.

Sheetz MP, Block SM, Spudich JA (1986). Myosin movement in vitro: a quantitative assay using oriented actin cables from Nitella. Methods Enzymol 134:531-544.

Trewavas AJ (1986). "Molecular and Cellular Aspects of Calcium in Plant Development". New York: Plenum Press.

Uyeda TQP, Furuya M (1987) ATP-induced movement between microfilaments and microtubules in myxomycete flagellates. Protoplasma 140:190-192.

Vasiliev JM (1987). Actin cortex and microtubular system in morphogenesis: Cooperation and competition. J Cell Sci 8(s): 1-18.

Wallace PJ, Packman CH, Wersto RP, Lichtman MA (1987). The effect of sulfhydryl inhibitors and cytochalasin on the cytoplasmic cytoskeletal actin of human neutrophils. J Cell Physiol 132:325-330.

Werz G (1968). Plasmatische Formbildung als Voraussetzung für die Zellwandbildung bei der Morphogenese von Acetabularia. Protoplasma 65:81-96.

Werz G (1969a). Wirkung von Colchicin auf die Morphogenese von Acetabularia. Protoplasma 67:67-78.
Werz G (1969b). Morphogenetic processes in Acetabularia. In Bücher T, Sies H. (eds): "Inhibitors - Tools in Cell Research", Berlin: Springer Verlag, pp 167-186.
Werz G (1970). Mechanism of cell wall formation in Acetabularia. In Brachet J, Bonotto S (eds): "Biology of Acetabularia", New York: Academic Press, pp 125-144.
Williamson RE (1984). Calcium and the plant cytoskeleton. Plant, Cell & Environm 7:431-440.
Woodcock CLF (1971). The anchoring of nuclei by cytoplasmic microtubules in Acetabularia. J Cell Sci 8:611-621.
Zimmer B, Werz, G (1981). Cytoskeletal elements and their involvement in Polyphysa (Acetabularia) protoplast differentiation. Cytoskeleton modifiers and concanavalin A-mediated effects. Exptl Cell Res 131:105-113.

Algae as Experimental Systems pages 93–108
© **1989 Alan R. Liss, Inc.**

THE ROLE OF THE CYTOSKELETON DURING THE ASSEMBLY, SECRETION,
AND DEPLOYMENT OF SCALES AND SPINES

Richard Wetherbee, Anthony Koutoulis and
Peter L. Beech
School of Botany, University of Melbourne,
Parkville 3052, Victoria, Australia

Much of the progress in our current understanding of
cell biology comes from intensive studies of relatively
simple systems, particularly unicellular organisms. Histor-
ically, microalgae have provided cell biologists with model
systems for investigating a number of processes, including
the role of the cytoskeleton in generating cellular and
subcellular movement and in determining cell shape. Similarly, scale-bearing algae have proven useful in studying
the formation and development of cell surface components,
particularly the role of the endomembrane system in the
synthesis and secretion of scales. In several chrysophycean algae, which are adorned with silicious scales,
cytoskeletal components closely associate with the scale
forming vesicles and appear to function in several different ways (Mignot & Brugerolle, 1982; Brugerolle & Bricheux,
1984). Microtubules and microfilaments help shape the
microarchitecture of the highly ornate scales/spines,
generate their precise movements prior to secretion and
affect their deployment onto the surface. In unicells
of Apedinella radians (Pedinellales, Chrysophyceae), a
species possessing spine-scales longer than the cell, the
cytoskeleton has a more spectacular role in generating and
coordinating the instantaneous reorientation of the elongate spine-scales on the cell surface (Koutoulis et al.,
1988). It is these and other interesting phytoflagellates
that are the subject of this review, as they are proving
ideal systems for studying the structure and function of
several cytoskeletal components as well as their
interactions during all stages of cell surface development.

SCALES AND SCALE-BEARING ALGAE

Although many scale types are observed by light microscopy, they were brought to the attention of cell biologists in the mid-fifties (Asmund, 1955; Fott,955; Manton,
1955; Parke et al., 1955) when the first electron micrographs of isolated scales were published. Manton and
Leedale (1961) subsequently observed that the scales were
formed within cells and then deposited onto the surface.
These investigations first demonstrated the value of scale-
bearing algae for studying the structure and function of
the Golgi apparatus. As most scales are highly ornamented,
and therefore identifiable on the surface and within cytoplasmic scale-forming vesicles, their sequential assembly
was followed from static electron microscopical images
taken of cells in key stages of development.

A large number of scale-bearing species from several
classes of algae have been investigated in detail (see
Romanovicz, 1981), mainly due to the usefulness of scale
morphology as a taxonomic tool; scale microarchitecture is
specific for a single species. In general, algal scales
are composed of polysaccharide, including cellulose in some
cases, with a small amount of protein (e.g. Romanovicz &
Brown, 1976). Some scale types become mineralized, either
with calcium carbonate (coccoliths) or silica. The presence of scales among the algae was originally thought confined to the green algae, particularly the Prasinophyceae,
and to the Chrysophyta, including the classes Chryso-
phyceae, Synurophyceae and Prymnesiophyceae. More recently, scales have been observed on the surface of a few
dinoflagellate genera and on the flagella and cell surface
of several cryptomonads (Pennick, 1981; Brett & Wetherbee,
1986). Individual species often contain more than one
scale type, with representatives from the prasinophyte
genus Pyramimonas having up to six different scale types,
elaborately stacked and positioned on the flagella and cell
surface (Moestrup & Walne, 1979; McFadden & Wetherbee,
1982). With the possible exception of the cryptophytes
(Wetherbee et al., 1986), scales are always synthesized and
pre-packaged within the endomembrane system prior to
secretion (Romanovicz, 1981). However, the sites of
formation and secretion can vary between groups.

Cytoskeletal components have not been observed
associated with scale formation and secretion in most

algae, though it is possible they have gone unnoticed, as individual microtubules and small numbers of microfilaments can be difficult to detect in electron micrographs. One exception is the presence of actin microfilaments and microtubules during the formation of silicious scales and spines in Synura and Mallomonas (Synurophyceae) (Brugerolle & Bricheux, 1984; Leadbeater, 1984,1986), organisms we will consider in some detail.

SCALE, BRISTLE FORMATION IN SYNURA AND MALLOMONAS

One of the main criteria used by Andersen (1987) for the creation of the new class Synurophyceae was the site and mechanism of silicious scale/spine formation. The common genera Synura and Mallomonas possess heavily silicified scales/spines (Fig.1-3,6-8). The formation of the latter occurs on the outer surface of one chloroplast only, (Schnepf & Deichgraber, 1969; Wujek & Kristiansen, 1978; Mignot & Brugerolle, 1982; Leadbeater, 1986). The silica deposition vesicle (SDV) reportedly arises from the fusion of several small vesicles derived from the Golgi and transported to the site of scale formation by an array of microtubules that remains associated with, but not connected to, the distal surface of the SDV. These microtubules apparently continue to position the SDVs during their developmental migration and, in addition, may also have a structural role in stabilizing the membrane prior to silica deposition and in helping to mould the scales (Fig. 4,5). As the SDV increases in size, forming the mould for the synthesis of the new scale, actin microfilaments attach directly to the outer membrane of the SDV adjacent to the chloroplast and develop into a cross-linked horseshoe pattern that underlies and predicts the position of the future scale base plate. Microfilaments may be absent from the central region of the SDV membrane if the periplastial ER is attached to it at this point. In such cases the periplastial ER eventually protrudes into the SDV to form a diverticulum which molds the hull or spine of the scale in species that have this scale type. The tenure of the microfilaments is short-lived, and they disappear soon after silica deposition begins. The microfilaments therefore appear to have a structural role in strengthening and shaping the SDV membrane until there is sufficient silica deposition to take over this role. The microfilaments may also be involved in molding the

microarchitecture of the ornate scales (Mignot &
Brugerolle, 1982; Brugerolle & Bricheux, 1984).

The mechanism of frustule morphogenesis in the diatoms
is similar in many ways to the formation of silica scales
(see Pickett-Heaps et al., 1988). In both instances it is
not clear whether the SDV is derived directly from the
Golgi, although presumably the SDV has the same basic
properties as a Golgi cisterna: it must contain the
biochemical machinery to concentrate and deposit silica
during scale or frustule formation, and presumably the
information for the precise, species-specific patterning of
these wall components (Pickett-Heaps et al, 1988). It is
also possible that the cytoskeletal components closely
associated with the SDV may have a role in mediating the
morphogenesis of the specific wall patterns that
characterize individual species of diatoms and
chrysophytes. How the genetic information is received and
processed by the membrane is not known.

BRISTLE DEPLOYMENT IN MALLOMONAS SPLENDENS

Spine formation and secretion is relatively well
understood in Mallomonas (Mignot & Brugerolle, 1982;
Brugerolle & Bricheux, 1984), though several interesting

Fig. 1 Silicious scales of Synura petersenii (x 3000).
Fig. 2 Scales and bristles of Mallomonas papillosa (x
2800). Each scale bears one bristle. Fig. 3 Single scale
of Mallomonas striata with attached bristle (x 9000). Fig.
4 Thin section of M. splendens showing body scale
formation in a SDV (open arrow) associated with the
periplastidial ER, microfilaments (arrows) and microtubules
(arrowheads) (x 25,000). Fig. 5 Bristle formation in SDV
(open arrow) of M. splendens with associated
microfilaments (arrows) and microtubules (arrowheads) (x
47,000). Fig. 6 - 8 Scanning EMs of M. splendens. Fig. 6
Cell in interphase showing the body scales, 4 large
bristles at either end attached to base-plate scales and
the flagellum (F) (x 1300). Fig. 7 Posterior end of a
pre-division cell with 8 base-plate scales and bristles (x
3400). Fig. 8 Cell in mid-cytokinesis with 8 posterior
bristles. Each developing daughter cell has only 2 anterior
bristles but 4 anterior BPSs (arrows) (x 1200).

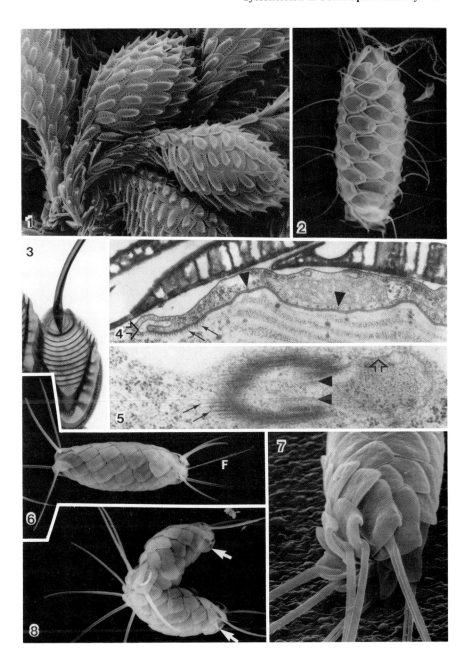

questions concerning the deployment of spines, or bristles
as they are called in this genus, have not been addressed.
M. splendens provides a particularly good system for
studying bristle deployment, as cells are covered with
silica scales and, at interphase, four anterior and four
posterior silica bristles (Fig. 6). Just prior to mitosis
the four posterior bristles double in number (Fig.7), but
the anterior bristles are doubled only after cytokinesis is
complete (Fig.8). When cells are resting, or if they are
preserved by fixatives, the bristles protrude at an oblique
angle to the long axis of the cell (Fig. 10). However,
when the cells are motile the anterior bristles are folded
back along the cell and the posterior bristles trail behind
(Fig. 9), presumably due to the hydrodynamic forces
involved in movement through water. Each bristle is
attached at its base to a specialized body scale, the base
plate scale (BPS), by a system of fibrils (Fig. 11) which
must be somewhat elastic as the bristles reorient during
swimming. The BPSs are formed separately from the bristles
and secreted after the bristles are already on the surface;
how they come together is presently not known.

Once a bristle is fully formed, the SDV fuses with the
plasma membrane and the bristle is deposited onto the
surface of the plasma membrane, but beneath the overlying
scale layer (Fig. 12 - 14). The base of the bristle is
attached to the plasma membrane by the same type of
fibrillar material that eventually attaches the bristle to
the BPS, while the remainder of the bristle is free (Fig.
12-14). The two sets of four bristles destined for the
anterior and posterior ends of the cell are deployed by a
different mechanism and at different times. Prior to
mitosis, four new posterior bristles are secreted beneath
the scale layer and then extruded, base first, out of the
cell (Fig. 15-18). Accompanying the extruding bristles is
a transient, cytoplasmic protuberance that emerges from the
cell posterior and grows outward with the bristles (Fig.
17-24). This tentacle-like structure remains relatively

Fig. 9, 10 Light micrographs of M. splendens. See text
for discussion (x 625). Fig. 11 - 14 Thin sections of M.
splendens. Fig. 11 Fibrillar attachment site between
bristle and base-plate scale (arrowheads) (x 25,000). Fig.
12 Posterior of cell showing bristles after secretion but
before extrusion and deployment. Note their location

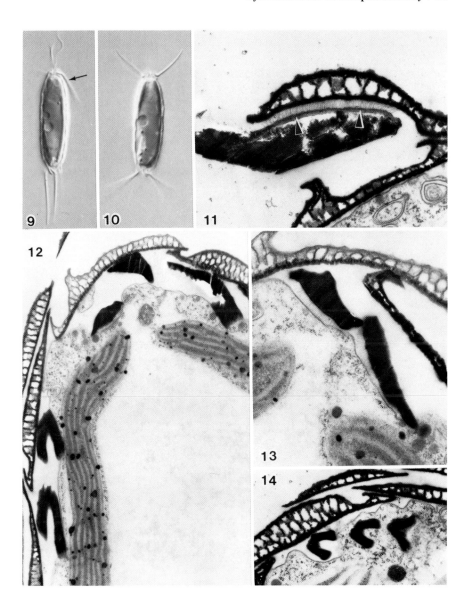

beneath the scale layer (x 16,500). Fig. 13 Bristle bases are connected to the plasma membrane by fibrillar material following secretion (x 27,000). Fig. 14 Three maturing bristles in cross-section; one has been secreted to the exterior and lies beneath the body scales (x 14,000).

stiff and the bases of the bristles appear attached to it
(Fig. 19,20). It is conceivable, even likely, that this
structure is controlling the movement of the bristles
during their subsequent reorientation, and is seen best in
image enhanced light microscopy. When the bristles are
fully extruded, a process that takes 5-8 minutes to
complete, and the tips are free of the cell, the base of
each bristle is drawn back to the posterior in a
relatively quick action (.5 - 2 mins; Fig. 19-22). The
bristle bases remain associated with the cytoplasmic
protuberance, which has also largely retracted by this
stage, but seem to be flexibly attached to the cell
posterior since the distal ends of the bristles drift
around prior to assuming their final orientation (Fig.
22,23). Prior to BPS secretion, the cytoplasmic
protuberance is completely retracted and the four new
posterior bristles are intimately associated at their basal
ends to the very posterior apex of the scaly cell covering
(Fig. 25,26).

Anterior bristles are extruded after cytokinesis is
complete. Each daughter cell adds two new bristles to the
two retained from the parent cell. These bristles are
formed in the same orientation as the new posterior
bristles and are therefore extruded tip first.
Reorientation is not required and the bristles join up with
the BPSs which are already in place. Our studies on these
cells are only beginning, but already this system shows
great promise for studying cellular control over the
deployment of cell surface components.

SPINE-SCALE REORIENTATION IN <u>APEDINELLA SPINIFERA</u>

Cells of this unusual chrysophyte possess scales and
six elongate spines (or spine-scales as they are called in
this organism) that are longer than the cells that produce
them. Motile cells have the ability to instantaneously
reorient the six spines while on the cell surface.
When swimming forwards the six spines normally project
posteriorly, forming a cone behind the cell (Fig.27).
However, upon a certain unknown stimulus, the spines can
alter their orientation, with no noticeable delay, to
project laterally. Following the reorientation, the spines
return (relatively more slowly) to their posterior
orientation. Although spine reorientation is accompanied

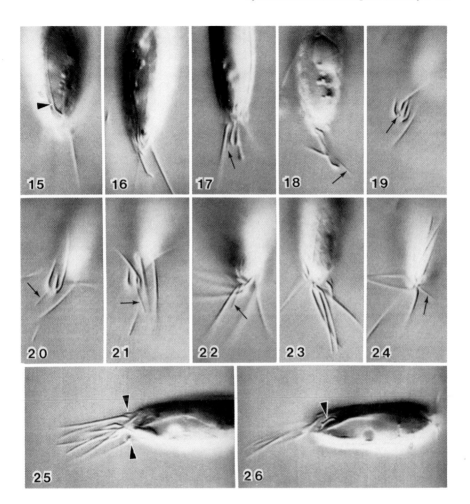

Fig. 15 - 26 Image enhanced light micrographs of posterior
bristle formation (x 750). Bristles are being extruded in
figures 15 - 18 while retraction of the bristle bases back
to the cell and subsequent bristle reorientation occurs in
figures 19 - 24. The tubular protuberance is indicated by
the arrow. Figure 25 is after reorientation, the 4 new
bristles are attached at a single point on the cell surface
while two parental base-plate scales and bristles are
evident (arrowheads). In another plane of focus seen in
figure 26, a base-plate scale (arrowhead) is about to be
secreted and meet up with one of the new bristles.

by a change of direction, cells also have the ability to
change directions without such an event.

The detailed structure of these cells has recently
been published (Koutoulis et al., 1988) and will only be
summarized here. Each spine is connected to the plasma
membrane by a cross striated fibrous band (termed a
microligament) (Fig. 28-31). Multi-layered plaques are
located beneath the plasma membrane at the attachment sites
(Fig. 29,30,33) and are connected to an extensive fibrous
system which appears involved in spine-scale reorientation
(Fig. 33). When spine-scales project posteriorly, plaques
are located on the circumference of the cell (Fig. 29).
When they project laterally the plaques are located some
distance from the circumference between the chloroplasts,
and consequently the microligaments are drawn into deep
invaginations (Fig. 30,33,35). No significant
difference in the length of the microligament or the
periodicity of its cross-striations are observed when the
spines are in the two different orientations. We therefore
presume that microligaments act merely to link the spine-
scales to the cell surface.

Immunocytochemistry in conjunction with sectioned
material reveals a complex cytoskeleton containing at least
three varieties of elongate elements: microtubules (Fig.
34,35,36,37), actin microfilaments (Fig. 33,38) and
centrin filaments (Fig. 35,39,40) (Salisbury et al., 1984).

Fig. 27   Cell of Apedinella radians with single flagellum
(arrow) and spine-scales trailing behind (x 1500). Fig. 28
Triangular base of a spine-scale with attached
microligament (arrow) (x 11,500). Fig. 29, 30 Two
positions of the plaques (PL) during spine-scale reorientation, either on the surface or drawn into the cell. Note
the striated microligaments (x 65,500; x 36,200). Fig. 31
Type d microtubular triad within a tentacle closely aligned
with a spine-scale. The microligament is in glancing
section (arrow) (x 23,600). Fig. 32 Transverse section of
a tentacle with type d microtubular triad and associated
spine-scale (x 30,000). Fig. 33 Transverse section
through the filamentous network responsible for spine-scale
reorientation. The filamentous bundles are anchored at the
plaques and the cylindrical caps (arrow) (x 20,200). Fig.
27-33 from Koutoulis et al., 1988.

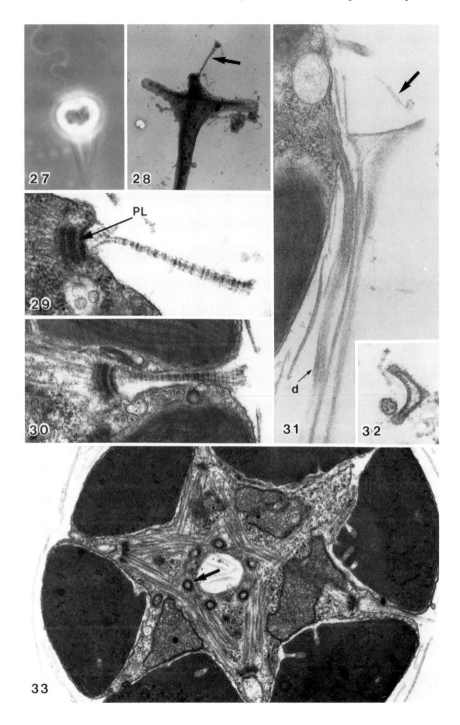

All three components interconnect at two distinct anchoring
sites, the plaques and a series of twelve cylindrical caps
(see below), to form an intricate three-dimensional network
(Fig. 33-35). Cytoskeletal microtubules are mostly found
as triads of three parallel microtubules arranged in an
equilateral triangle interconnected by thin filaments (Fig.
31,32). All triads emanate from the nuclear membrane
and thirty-one are present in each cell during interphase.
Because the cells display radial symmetry, the microtubules
can be separated into five distinct types (see diagram,
Fig. 34): Type a) 12 triads project anteriorly and are
surrounded at their apex by hollow, electron dense
cylindrical caps (CCs; Fig. 33) and further divided into
two whorls of 6 inner CCs and 6 outer CCs, with the outer
whorl forming an anterior tier and the inner whorl being
posterior to it. Type b) 6 triads project anterio-
laterally and continue externally to form tentacles above
the plaques, lying close to the microligaments and
apparently lying above the spines during swimming. Type c)
6 triads project anterio-laterally and then curve
posteriorly between the six chloroplasts where they meet a
sphincter at the posterior end of the cell. Type d) 6
triads project anterio-laterally and then continue
externally just beneath the plaques to form tentacles that
stay closely associated to the proximal surface of the
spine whether they are oriented posteriorly or laterally
(Fig. 31,32,34);

Fig. 34,35 Diagrams of the cytoskeleton of A. radians in
longitudinal and transverse view. See text for discussion.
ML = microligament; FL = flagellum; PY = pyrenoid; CH =
chloroplast; CC = cylindrical caps; iCC = inner whorl of
cylindrical caps; oCC = outer whorl of cylindrical caps; PL
= plaques; SS = spine-scale; BS = body scales; TS =
trailing stalk; TE = tentacle; SP = sphincter; ao = outer
type a triads; ai = inner type a triads; b,c,d,d =
microtubular triads; actin bundles = arrowheads; centrin
bundles = arrows. Fig. 36,37 Fluorescent micrographs
using FITC labelled anti-tubulin, and showing the presence
of the triads (x 1800). Fig. 38 FITC labelled anti-actin
antibody revealing the filamentous bundles of actin (x
1800). Fig. 39,40 FITC labelled anti-centrin reveals a
central staining region associated with the basal bodies
(arrowhead) plus the filamentous bundles (arrow) (x 1800).
Fig. 34-40 from Koutoulis et al., 1988.

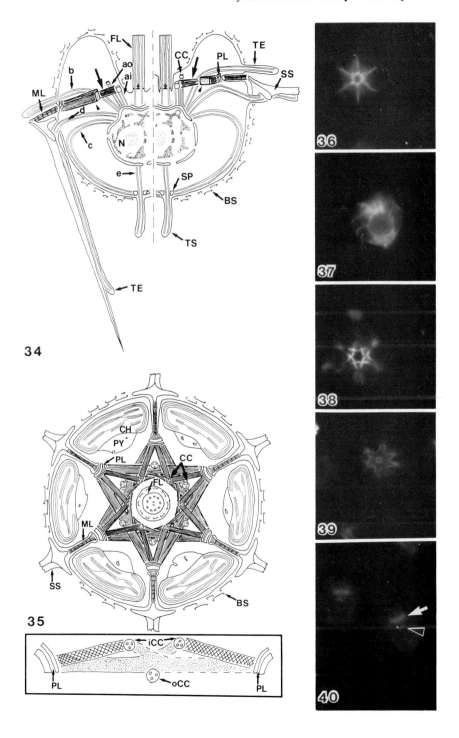

attached to the spines, they probably have a function in orienting the spines during movement. Type e) 1 triad projects posteriorly from the nucleus, passes through the sphincter and supports a trailing stalk (Fig 34).

A complex filamentous network occurs in the anterior region of the cell and is composed of at least actin and centrin in the form of filamentous bundles (Fig. 33 - 35) (Koutoulis et al., 1988). Using immunofluorescent techniques and sectioned material, we have been able to determine the structure and positioning of the two different types of filamentous bundles. Two bundles of actin microfilaments emanate from each plaque and attach to every alternate plaque, thereby forming two overlapping, triangular networks reminiscent of the "Star of David" (Fig. 35). Additionally, each actin filamentous bundle is branched; the branch attaches itself to one of the innermost whorls of cylindrical caps, thereby establishing both plaque/plaque and plaque/CC connections (Fig. 35, inset). Two centrin filamentous bundles emanate from each plaque and attach to the two closest members of the innermost whorl of CCs. A second, quite different, six-pointed star results from plague/CC connections only (Fig. 35,39,40). Additionally, centrin appears to surround the base of the two basal bodies at a position adjacent to the apical pole of the nucleus, a situation described for a number of algal flagellates (Salisbury et al., 1984).

We proposed that the contraction of the actin and/or centrin filamentous bundles is responsible for pulling the plaques into the deep invaginations and consequently reorienting the spines (Koutoulis et al., 1988). The type d tentacles are believed to coordinate spine movement and to keep them evenly spaced during reorientation. It is possible that they may also lift the spines in some fashion, though this would seem unlikely when considering the extensive filamentous network which appears better positioned to undertake this role.

CONCLUSIONS

The microalgae discussed in this review provide excellent systems for investigating the interactions between the various components of the cytoskeleton as well as its role in a variety of subcellular, cellular and extracellular processes. Of particular interest is the external coordination achieved by cells during the deployment and subsequent movement of spines on the cell surface. These cells are able to carryout a number of unusual activities beyond the bounds of the cell membrane, and the discovery of the mechanisms and responsible molecules promises to be an exciting field of investigation during the next few years, with microalgae once again providing the necessary, simple systems.

REFERENCES

Asmund B (1955). Electron microscope observations of Mallomonas caudata and some remarks on its occurrence in four Danish Lakes. I Bot Tidskr 52:163-172.
Brett SJ, Wetherbee R (1986). A comparative study of periplast structure in Cryptomonas cryophila and C. ovata (Cryptophyceae). Protoplasma 131:23-31.
Brugerolle G, Bricheux G (1984). Actin microfilaments are involved in scale formation of the chrysomonad cell Synura. Protoplasma 123:203-212.
Fott IB (1955). Scales of Mallomonas observed in the electron microscope. Preslia 27:280-287.
Koutoulis A, McFadden GI, Wetherbee R (1988). Spine-scale reorientation in Apedinella radians (Pedinellales), Chryso-phyceae): the microarchitecture and immunocytochemistry of the associated cytoskeleton. Protoplasma (in press).
Leadbeater BSC (1984). Silicification of cell walls of certain protistan flagellates. Phil Trans R Soc Lond B 304:529-536.
Leadbeater BSC (1986). Scale case construction in Synura petersenii Korsch (Chrysophyceae). In Kristiansen J, Andersen RA (eds): "Chrysophytes: aspects and perspectives," Cambridge: Cambridge University Press, pp 121-131.
Manton I (1955). Observations with the electron microscope on Synura caroliniana Whitford. Proc Leeds Phil Lit Soc 6:306-316.

Manton I, Leedale G (1961). Observations on the fine
    structure of Paraphysomonas vestita with special
    reference to the Golgi apparatus and the origin of
    scales. Phycologia 1:37-57.
McFadden GI, Wetherbee R (1982). An investigation of the
    periplast scale layers in the antarctic flagellate
    Pyramimonas gelidicola (Prasinophyceae). Micron 13:329-
    330.
Mignot JP, Brugerolle G (1982). Scale formation in the
    Chryso-monad flagellates. J Ultrastruct Res 81:13-26.
Moestrup O, Walne PL (1979). Studies on scale
    morphogenesis in the Golgi apparatus of Pyramimonas
    tetrarhynchus (Prasinophyceae). J Cell Sci 36:437-459.
Parke M, Manton I, Clarke B (1955). Studies on marine
    flagellates. II. Three new species of Chrysochromulina.
    J mar biol Ass UK 34:579-609.
Pennick NC (1981). Flagellar scales in Hemiselmis
    brunnescens Butcher and H. virescens Droop. Arch
    Protistenk 124:267-270.
Pickett-Heaps JD, Schmid A-MM, Edgar LE (1988). Cell
    biology of diatom valve morphogenesis. Progress in
    Protistology (in press).
Romanovicz DK (1981). Scale formation in flagellates. In
    Kiermayer O (ed): "Cell Biology Monographs, Vol 8,
    Cytomorphogenesis in Plants," New York: Springer-Verlag,
    pp 27-62.
Romanovicz DK, Brown RM (1976). Biogenesis and structure
    of Golgi-derived cellulosic scales in Pleurochrysis. II.
    Scale composition and supramolecular structure. Appl
    Polymer Symp 28:587-610.
Salisbury JL, Baron AT, Surek B, Melkonian M (1984).
    Striated flagellar roots: isolation and partial
    characterization of a calcium-modulated contractile
    organelle. J Cell Biol 99: 926-970.
Schnepf E, Deichgraber G (1969). Uber die feinstruktur
    von Synura petersenii unter besonderer berucksichtigung
    der morphogenese ihrer kieselschuppen. Protoplasma
    68:85-106.
Wetherbee R, Hill DRA, McFadden GI (1986). Periplast
    structure of the Cryptomonad flagellate Hemiselmis
    brunnescens. Protoplasma 131:11-22.
Wujek DE, Kristiansen J (1978). Observations of bristle
    and scale production in Mallomonas caudata
    (Chrysophyceae). Arch Protistenk 120:213-221.

Section II.
Morphogenesis and Differentiation

Algae as Experimental Systems pages 111–119
© 1989 Alan R. Liss, Inc.

POLARIZATION IN *FUCUS* (PHAEOPHYCEAE) ZYGOTES: INVESTIGATIONS
OF THE ROLE OF CALCIUM, MICROFILAMENTS AND CELL WALL

Ralph S. Quatrano[1] and Darryl L. Kropf[2]

Department of Botany/Plant Pathology
Oregon State University
Corvallis, Oregon 97331

Brown algae of the order Fucales have been used for almost 100 years
as a model to study the mechanism by which environmental gradients
impose a polarity on the early development of an organism (Whitaker,
1940; Jaffe, 1968; Evans, et al. 1982; and Quatrano, et al. 1985).
Zygotes and young embryos of the Fucales also provide a model system to
study the events of early embryogenesis in plants; egg activation and cell
wall formation following fertilization, and the generation of a common
morphogenetic and segmentation pattern in the young embryo. The
pattern of early morphogenesis in *Fucus* is remarkably similar to that of
other plant groups; an unequal first division of the zygote resulting in two
cells whose lineages give rise to different structures (c.f. Wardlaw,
1968). All the processes of early embryo development are difficult to
study and experimentally manipulate in higher plants because the
fertilized eggs are deeply buried in ovule tissue and do not develop *in
vitro* after isolation. In contrast, eggs of the Fucales are released from
the mature plant in large numbers, with fertilization and subsequent
zygote and embryo development occurring synchronously in a defined
medium (Quatrano, 1980). This paper will focus on the role of various
factors in the establishment of a primary polar axis in the zygote of *Fucus*
and our most recent experiments aimed at understanding the polarization
process.

[1]To whom all correspondence should be addressed: Department of Biology,
   University of North Carolina, Chapel Hill, NC   27599-3280
[2]Department of Biology, University of Utah, Salt Lake City, UT   84112

Earlier studies demonstrated that eggs and young zygotes of *Fucus* are apolar, unlike the eggs of most animals or higher plants (cf. Jaffe, 1968; Quatrano, et al., 1979, Quatrano, et al., 1985). Moreover, the polar axis in *Fucus* can be oriented by a number of vectors, unilateral light being the most extensively and easily utilized. When zygotes are subjected to a gradient of light, the plane of the first cell division is always perpendicular to the light axis. The resulting cell plate divides two unequal cells, the smaller rhizoid cell on the shaded portion of the gradient and the thallus cell on the lighted side. Polarization of the developmental axis by light has two components; axis formation and axis fixation (Quatrano, 1973). An axis formed at any time during the first 10 h of development is labile and can be easily changed to a new direction by reorienting the zygote to a light pulse from a different direction. Between 10 and 12 h, the last axis formed will be fixed in space so that other orienting vectors are incapable of forming an axis. This fixed axis then directs the accumulation of components that results in germination (i.e. expression of rhizoid and thallus polarity) of the zygote at 14 h. The first cell division occurs at about 20 h producing the rhizoid and thallus cells (cf. Kropf & Quatrano, 1987 for discussion of the timing of the developmental program of fucoid zygotes). Our main interest is to determine the cytological and biochemical basis of axis fixation in this polarization process, and in particular, the role of calcium, microfilaments and the cell wall.

ROLE OF CALCIUM

Jaffe, Robinson and Nuccitelli (Robinson & Jaffe, 1975; Nuccitelli, 1978) elegantly demonstrated in a series of papers that an electrical current flows through the zygote during the polarization process; into the presumptive rhizoid pole and out of the presumptive thallus pole. The current begins to flow at the time of axis formation in *Pelvetia* (Fucales) but has not been measured in *Fucus*. The site of inward current precedes and accurately predicts the site of rhizoid outgrowth. Some part of this current is carried by Ca^{+2} and leads to the hypothesis that Ca^{+2} channels are localized at the presumptive rhizoid end of the light gradient, causing a net influx of Ca^{+2} into the "dark" end with a net efflux on the "light" or thallus end. Brownlee and Wood (1986), using Ca^{+2} selective electrodes, found the Ca^{+2} concentration to be 10x greater in the tip of the rhizoid cytoplasm than in the sub-tip region. The resulting Ca^{+2} gradient in the rhizoid is hypothesized to play a role in the formation of the polar developmental axis in the zygote. Two predictions essential for polarization arise from these studies; a demonstrable Ca^{+2} gradient in the cytoplasm of the zygote, and a directed Ca^{+2} influx into the zygotes.

The fluorescence probe chlorotetracycline (CTC) binds Ca^{+2} in a membranous environment but does not measure free Ca^{+2} gradients. It is a convenient tool, however, especially since Ca^{+2} has been clearly localized in the growing tip of other systems (e.g., pollen and hyphae in Reiss & Herth (1979); moss tip cells in Saunders & Hepler (1981). When used in *Fucus*, CTC does not fluoresce in eggs, but exhibits uniform fluorescence in the zygotes at the time of polarization, and is clearly localized in the emerging rhizoid, at the tip of an elongating rhizoid (Fig. 1), as well as in all cross walls of the young embryo (Kropf & Quatrano, 1987). However, about 20% of the cells did exhibit

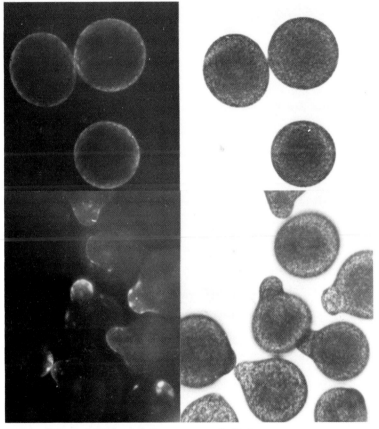

Figure 1 Phase (right) and fluorescent (left) images of 6 h zygotes (top) and two-celled embryos (bottom) stained with CTC showing symmetrical localization of membrane-bound Ca^{+2} in zygotes with a fixed polar axis (top) as compared to a localized accumulation of Ca^{+2} in the tip of the elongating rhizoid (bottom).

asymmetric localization of CTC in the zygote about the time of axis formation, but it was not clear what the relationship of this site of localization was to the developmental axis. Using photopolarization to measure when axis fixation occurred (Quatrano, 1973), we were able to demonstrate clearly that Ca^{+2} localization (as measured by CTC fluorescence) was symmetrically distributed in all cells of a population 14 h after fertilization, i.e. after axis fixation (Kropf & Quatrano, 1987). Hence, these data do not support the hypothesis that the pool of Ca^{+2} detected by CTC plays a role in polarization.

Using ethyleneglycol-bis-(β-aminoethyl ether-N,N,N′N′-tetracetic acid (EGTA) as a specific chelator of extracellular Ca^{+2}, we also demonstrated that from 10^{-10} to 10^{-3}M, extracellular Ca^{+2} has no effect on the ability of cells to fix a polar axis formed by light. The Ca^{+2} channel blockers, lanthanum, D-600 and verapamil had no effect on this process, nor did the anion channel blocker DIDS (Kropf & Quatrano, 1987). Although Ca^{+2} localization and Ca^{+2} influx are not apparently required for polarization, they are clearly required for the tip-growth of the rhizoid, similar to tip-growing pollen tubes (Picton & Steer, 1983). To further our understanding of the role of Ca^{+2}, two experiments would be helpful; measurement of the free cytoplasmic Ca^{+2} in *Fucus* and *Pelvetia* in relation to the photopolarization process, and, the timing of the current flow relative to polarization in *Fucus*.

ROLE OF MICROFILAMENTS

Previous work demonstrated that: (1) microfilament inhibitors, (i.e. the cytochalasins), blocked axis formation and fixation (Quatrano, 1973; Brawley & Quatrano, 1979), and affected endogenous currents (Brawley & Robinson, 1985); (2) microfilaments (actin) are localized at or just before germination (Brawley & Robinson, 1985). Phalloidin, which specifically binds to filamentous actin (F-actin), was fluorescently tagged with rhodamine (Wulf, et al., 1979), and utilized in our studies to investigate the distribution of microfilaments during early development . Phalloidin detected F-actin symmetrically distributed in the cortical region of zygotes that do not have a fixed polar axis. Individual actin filaments were not resolved as in higher plants (cf. Lloyd, 1988), but a rather diffuse stain was observed that is probably indicative of a fine network of microfilaments, similar to that demonstrated in fungi (Hoch & Staples, 1983). Cells incubated in potassium iodide, which depolymerizes microfilaments, or pretreatment of cells with excess, unlabelled phalloidin were unstained. However, if rhodamine-phalloidin was added to ungerminated zygotes with a fixed polar axis (12 h after fertilization), a localized accumulation of F-actin was observed. A similar localization was recorded at the tip of two-celled or older embryos (Fig. 2 & cf. Brawley & Robinson, 1985). Using the same approach as with the CTC staining, we determined that the future site of rhizoid formation in fixed embryos is the area that stains preferentially with rhodamine-phalloidin. Furthermore, the timing of the F-actin localization coincides with the fixation process and not with the germination process as does localization of CTC staining. In the presence of cytochalasin, which prevents axis fixation, F-actin staining is still present, but the asymmetric localization of the stain is not observed. Hence, the redistribution of F-actin to the subsequent site of rhizoid formation is temporally correlated with the fixation process and clearly implicates microfilaments in the process of axis fixation. Two possible mechanisms may account for the redistribution of F-actin: 1) a depolymerization at the thallus end and a repolymerization at the rhizoid end; 2) cortical flow of actin (cf. Bray & White, 1988) towards the rhizoid end.

Figure 2 Fluorescent images of 12 hour zygotes with a fixed polar axis (left) and two-celled embryos (right) stained with rhodamine-phalloidin showing the localized accumulation of F-actin at the site of rhizoid formation.

Finally, the amount of actin present in various stages of embryo development, and the timing of its biosynthesis were determined by extracting protein, electrophoretically separating the three isoforms by two-dimensional gel electrophoresis and subjecting the separated proteins to immunoblot (using monoclonal antibody to chicken gizzard actin from Amersham) and fluorography (after incubation of zygotes in ^{14}C-labeled bicarbonate). Although the concentration of actin present in the eggs, zygotes and young embryos does not change, actin synthesis is triggered in zygotes soon after fertilization and continues to be synthesized throughout development. No synthesis was detected in eggs (Fig. 3). The time of protein synthesis required for axis fixation (prior to 6 h - Quatrano, 1968) is the same as when actin synthesis is initiated, and suggests further a close association between actin and axis fixation (cf. Quatrano, et al., 1979; Brawley & Robinson, 1985).

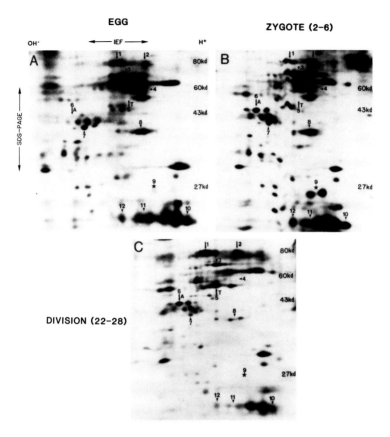

Figure 3 Two dimensional fluorographs of proteins synthesized at various stages of early *Fucus* development. Three actin isoforms (6A) are not synthesized in eggs (A) but are clearly detectable in 2-6 h zygotes (B) and in 22-28 h two-celled embryos (C).

ROLE OF CELL WALLS

We have recently shown that protoplasts prepared from young zygotes can regenerate a cell wall and continue to develop into polarized two-celled embryos (Kropf, et al., 1988). The questions we asked using this system were: in the absence of a cell wall; 1) can protoplasts subjected to unilateral light form an axis?; 2) can protoplasts fix a previously formed axis? Protoplasts were given a 6 h unilateral light pulse (during which time no cell wall regeneration occurred due to the presence of a diluted enzyme solution), which resulted in 86% of the regenerated embryos forming a rhizoid cell on the shaded side of the light gradient (Table 1(C), in Kropf, et al., 1988). Using a light gradient from a second direction, still while the protoplasts were prevented from forming a wall, we observed that the polar axis remained labile in the absence of a cell wall (Table 1(D), in Kropf, et al., 1988). Hence, the presence of a cell wall is an absolute requirement for axis fixation.

SUMMARY

Our working hypothesis at present is that axis fixation involves transmembrane bridges at the presumptive rhizoid pole, from the cell wall to the microfilament cytoskeleton. These structural components, both having been shown above to be required for axis fixation, anchor Ca^{+2} channels in the plasma membrane. The Ca^{+2} channels accumulate at the presumptive rhizoid site in response to a light gradient by an unknown mechanism. Calcium localization in membranous compartments appears not to be associated with axis fixation but rather with rhizoid tip growth. The roles of free Ca^{+2} and its localization in the zygotes, as well as the role of the endogenous electrical current are still unknown.

ACKNOWLEDGEMENTS

Our research reported in this paper was supported by an NSF Post-doctoral Fellowship in Plant Biology (DCB 8412389) to DLK and an NSF grant (DCB 851152) to RSQ. We thank Roswitha Hopkins and Sonja Berge for their technical assistance and Jeanie Griffiths for help in preparing the manuscript.

REFERENCES

Brawley SH, Quatrano RS (1979). Sulfation of fucoidin in Fucus embryos IV Autoradiographic investigations of fucoidin sulfation and secretion during differentiation and the effect of cytochalasin treatment. Devel Biol 73:193-205.
Brawley SH, Robinson KR (1985). Cytochalsasin treatment disrupts the endogenous current sassociated with cell polarization in fucoid zygotes: studies of the role of F-actin in embryogenesis. J Cell Biol 100:1173-1184.
Bray D, White JG (1988). Cortical flow in animal cells. Science 239:883-887.

Brownlee C, Wood JW (1986). A gradient of cytoplasmic free calcium in growing rhizoid cells of Fucus serratus. Nature 320:624-626.

Evans LV, Callow JA, Callow ME (1982). The biology and biochemistry of reproduction and early development in Fucus. Prog Phycol Res 1:67-110.

Hoch HC, Staples RC (1983). Visualization of actin in situ by rhodamine-conjugated phalloidin in the fungus Uromyces phaseoli. Eur J Cell Biol 32:52-58.

Jaffe LF (1968). Localization in the developing Fucus egg and the general role of localizing currents. Adv Morphog 7:295-328.

Kropf DL, Quatrano RS (1987). Localization of membrane-associated calcium during development of fucoid algae using chlorotetracycline. Planta 171:158-170.

Kropf DL, Kloareg B, Quatrano RS (1988). Cell wall is required for fixation of the embryonic axis in Fucus zygotes. Science 239:187-190.

Lloyd C (1988). Actin in plants. J Cell Sci 90:185-188.

Nuccitelli R (1978). Ooplasmic segregation and secretion in the Pelvetia egg is accompanied by a membrane-generated electrical current. Devel Biol 62:13-33.

Picton JM, Steer MW (1983). Evidence for the role of Ca2+ ions in tip extension in pollen tubes. Protoplasma 115:11-17.

Quatrano RS (1968). Rhizoid formation in Fucus zygotes: dependence on protein and ribonucleic acid synthesis. Science 162:468-470.

Quatrano RS (1973). Separation of processes associated with differentiation of two-celled Fucus embryos. Devel Biol 30:209-213.

Quatrano RS (1980). Gamete release, fertilization, and embryogenesis in the Fucales. In Gantt, E (ed): "Handbook of Phycological Methods: Developmental and Cytological Methods" Cambridge University Press, 59-68.

Quatrano RS, Brawley SH, Hogsett WE (1979). The control of the polar deposition of a sulfated polysaccharide in Fucus zygotes. In "Determinants of Spatial Organization". S. Subtelny, I. R. Konigsbert, editors. Academic Press, New York. 77-96.

Quatrano RS, Griffing LR, Huber-Walchli, V, Doubet, S (1985). Cytological and biochemical requirements for the establishment of a polar cell. J Cell Sci, Suppl 2:129-141.

Reiss HD, Herth W (1979). Calcium gradients in tip growing plant cells visualized by chlorotetracycline fluorescence. Planta 146:615-621.

Robinson KR, Jaffe LF (1975). Polarizing fucoid eggs drive a calcium current through themselves. Science 187:70-72.

Saunders MJ, Hepler PK (1981). Localization of membrane-associated calcium following cytokinin treatment in Fuvaria using chlorotetracycline. Planta 152:272-281.

Wardlaw CW (1968). "Morphogenesis In Plants". Methuen & Co., London: 451 pp.

Whitaker DM (1940). Physical factors of growth. Growth Suppl 75-90.

Wulf E, Deboben A, Bautz FA, Faulstich H, Wieland T (1979). Fluorescent phallotoxin, a tool for the visualization of cellular actin. Proc Natl Acad Sci USA 76:4498-4502.

Algae as Experimental Systems pages 121–134
© 1989 Alan R. Liss, Inc.

CELLULAR MORPHOGENESIS IN THE FILAMENTOUS RED ALGA GRIFFITHSIA

Susan D. Waaland

Department of Biology, University of Puget Sound, Tacoma, Washington 98416

Filamentous red algae, such as those in the genera Griffithsia, Anotrichium, Antithamnion and Callithamnion, are excellent organisms in which to study the cellular control of morphogenesis for a number of reasons. 1) Their simple, branched, filamentous morphology allows ready access to individual cells for observation and manipulation. These cells are large enough to be easily manipulated under a dissecting microscope. In genera such as Antithamnion branching pattern is quite rigid, in Griffithsia however, the number of branches per plant can be altered by changing light intensity (Waaland & Cleland, 1972). 2) Addition of cells to these filaments occurs only at the apex although subapical cells may produce branch initials and adventitious rhizoids. Thus, there is a gradient in cell age moving basipetally along a filament (Dixon, 1973). 3) Since cells are totipotent, new plants may be started using single vegetative cells, short vegetative filaments and protoplasts (Lewis, 1909; Konrad-Hawkins, 1964; L'Hardy-Halos, 1971a,b; Duffield et al., 1972). The pattern of development from single cells can be analyzed. 4) The morphology of plants is relatively simple, however there are at least two distinctive types of somatic cells: upright, deeply colored shoot cells; and, prostrate, pale, adhesive rhizoid cells (Duffield et al., 1972; L'Hardy-Halos & Larpent, 1983). There are physiological and biochemical as well as morphological differences between these two types of cells. In rhizoids, only the apical cell elongates but in shoot cells both apical and intercalary cells elongate. The pattern of elongation in shoot cells is a localized band growth (Waaland et al., 1972; Waaland & Waaland, 1975). In addition, there is a difference between rhizoids and shoot cells in their response to unilateral light. Rhizoids are negatively phototropic whereas shoot cells grow toward unilateral light (L'Hardy-Halos, 1971a; Waaland et al., 1977). Many aspects of development in these organisms are of interest. In this paper, I shall discuss some intriguing lines of research that have developed from studies of one developmental sequence: cell repair by cell fusion.

CELL REGENERATION VS. CELL REPAIR BY CELL FUSION

In genera such as Griffithsia and Anotrichium, when a cell in the middle of a filament is killed one of two distinct developmental pathways may be initiated: regeneration or cell repair. If the dead cell is severed, regeneration is induced (Fig. 1A). The cell above the severed cell regenerates a new rhizoid cell at its base. The cell below the dead cell regenerates a new, dome-shaped shoot apical cell at its apex (L'Hardy-Halos, 1971b; Duffield et al., 1972; Waaland & Cleland, 1974). Thus from the original plant, two new plants are produced. Normally, the polarity of regeneration is very regular; a cell regenerates a rhizoid from its basal end and a shoot cell from its apical end. This polarity is not reversed by light or gravity; however, unusual regeneration patterns have been observed when cells are centrifuged or placed in an external current gradient (Schecter, 1934).

If a cell in the middle of a filament is killed but its cell wall is not severed, the process of cell repair by cell fusion is induced (Fig. 1B; L'Hardy-Halos, 1971a; Waaland & Cleland, 1974). In this instance, the cell above the dead cell produces a rhizoid, as it would during regeneration. However, the cell below the dead cell produces a specialized, elongate repair shoot cell. The rhizoid and repair shoot cell grow towards each other in the lumen of the dead cell, are attracted to one another and when they meet, fuse to form a single cell. The new cell expands laterally to fill the wall of the dead cell it replaces and becomes indistinguishable from that cell. Thus, during cell repair, two cells, a rhizoid and a repair shoot cell are produced, fuse and redifferentiate to form an intercalary shoot cell; this process restores the continuity of the original filament. The process of cell repair has been observed in at least six species of Griffithsia, species of Antithamnion and Anotrichium tenue (as G. tenuis) (L'Hardy-Halos, 1971a;

Waaland and Cleland, 1974; Waaland, 1975; Waaland, unpublished). Regeneration from excised filaments occurs in species of Callithamnion, however the process of cell repair (Westbrook, 1927). Studies of the cell repair process and its control have lead to several intriguing avenues of research. Four of these will be discussed here including: 1) investigation of the hormonal control of the cell repair; 2) analysis of unusual results of intraspecific somatic cell fusion; 3) probes of the nature of cell differentiation; and 4) investigation of cytoplasmic polarity and localized growth.

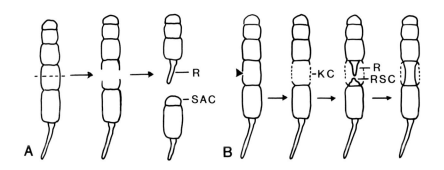

Figure 1: Cell Regeneration vs Cell Repair by Cell Fusion in Griffithsia: A. Regeneration after severing of a filament. B. Cell Repair after cell death without severing. KC=killed cell R= rhizoid, SAC= shoot apical cell, RSC= repair shoot cell.

HORMONAL CONTROL OF CELL REPAIR: CHARACTERIZATION OF RHODOMORPHIN

Several aspects of cell repair suggest that intercellular communication via a morphogenetic substance coordinates cell repair: 1) during cell repair, the cell below the dead cell divides much earlier than during regeneration (e.g., 6-8 h vs 16-18 h in G. pacifica (Waaland & Watson, 1980; Waaland, 1986); 2) the repair shoot cell, produced in cell repair differs in morphology and growth pattern from a shoot apical cell produced during regeneration (Waaland & Cleland, 1972; Waaland & Watson, 1980); and 3) there is a mutual attraction of rhizoid and repair shoot cell leading to cell fusion (Waaland & Cleland, 1972; Waaland, 1981). Evidence from in vivo and in vitro experiments has shown that rhizoids produce a species-specific substance that has been called "rhodomorphin" (Waaland & Cleland, 1972; Waaland, 1975; Waaland and Watson, 1980). Rhodomorphins induce early division of apices of decapitated filaments to produce repair shoot cells and are required for maintenance of elongation and morphogenesis in repair shoot cells.

Experimental evidence suggests rhodomorphins are produced in a number of species, however only rhodomorphin from G. pacifica has been isolated, purified and characterized (Waaland & Watson, 1980; Watson & Waaland, 1983, 1986). This molecule binds reversibly to Concanavalin A (Con A) therefore it contains carbohydrate with terminal α-mannosyl and/or α-glucosyl residues. If the molecule is treated with a-mannosidase, binding to Con A is reduced but activity is undiminished; this confirms the presence of α-mannosyl groups. That G. pacifica rhodomorphin contains protein is shown by the fact that activity is destroyed by heating the molecule to $50^{o}C$ for 8 min or by treatment with proteinases; treatment with sufhydryl reducing agents also destroys activity. Thus rhodomorphin from G. pacifica is a glycoprotein that contains a-mannosyl residues and essential sulfhydryl linkages; it has been shown to have a M_r 14-17,000 by gel filtration and SDS gel electrophoresis; it is active at 10^{-13}-10^{-14} M (Watson & Waaland, 1983, 1986).

Rhodomorphin from G. pacifica is the first endogenous morphogenetic hormone to have been characterized from red algae. It was only the second morphogenetic substance to be characterized from all algae; the first was the sexual inducer from the colonial green alga Volvox. The Volvox inducer is a basic glycoprotein with a molecular weight of about 30-32,000 (Kochert & Yates, 1974; Starr & Jaenicke, 1974; Starr & Jaenicke, this volume). It is intriguing that both of these substances, although quite different in chemical composition are glycoproteins; while glycoprotein hormones are common in animals their occurrence is rare among photosynthetic organisms.

DEVELOPMENTAL CONSEQUENCES OF INTRASPECIFIC SOMATIC CELL
FUSION

The process of cell repair involves fusion of two
somatic cells. Investigations of the developmental consequences of somatic cell fusion have been facilitated by the
development of an artificial in vitro cell repair system
(Fig. 2; Waaland, 1975). During cell repair in vivo, the
two halves of a filament are held together by the wall of
the dead cell; the wall serves two purposes: to contain and
concentrate rhodomorphin produced by the down-growing
rhizoid and to keep the two halves of the filament from
separating. In vitro, a substitute for the wall of the dead
cell must be found. In some species, e.g., Anotrichium
tenue, repair can be induced if a decapitated filament from
one plant is placed next to a rhizoid from another filament
(Waaland, 1978b). In other species, an enclosure is required. Glass capillary tubing does not work because filaments will not regenerate inside it. However satisfactory
tubing can be obtained by using cleaned cylinders of cell
walls from cells of green algae such as Nitella (Fig. 2;
Waaland 1975, 1978a; Waaland & Watson, 1980).

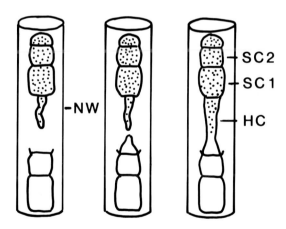

Figure 2. Artificial Cell Repair System in Nitella Wall
Cylinder: NW= cleaned cell wall, R= rhizoid, RSC= repair
shoot cell, HC= hybrid cell, SC-1= first shoot cell adjacent
to hybrid cell, SC-2= shoot cell above SC-1.

If a decapitated filament from one plant is inserted into
one end of the cylinder and a filament with a rhizoid is
inserted into the other end, the decapitated filament will
produce a repair shoot cell; the repair shoot cell and rhizoid will grow together and fuse to form a hybrid cell. In
vitro formation of intraspecific, somatic cell hybrids have
been produced between male and female plants, haploid and
diploid plants, and plants from different locales (Waaland,
1975; Koslowsky & Waaland, 1984). However, no interspecific
hybrids have been produced; the induction of repair shoot
cell formation, attraction of rhizoid and repair shoot cell,
and cell fusion appear to be species specific (Waaland,
1975).

In at least two cases, some unusual results have been
obtained after formation of somatic hybrids. The first
occurs following somatic cell fusion between two isolates of
G. pacifica, one (PP) from Puerto Peñasco, Mexico and the
other (BC) from Vancouver Island, Canada. Under in vitro
conditions, cell repair and production of a hybrid cell
between the two isolates proceeds normally. After a period
of several hours, however, an incompatibility reaction is
initiated in the hybrid cell (Waaland, 1975; Koslowsky &
Waaland, 1984). Chloroplasts from the PP parent form aggregates and fuse; over a period of several days the PP-chloro-
plasts degenerate. Ultrastructural analysis shows that
chloroplasts appear to be the only organelle affected; mitochondria and nuclei from the PP strain retain a normal appearance (Koslowsky & Waaland, 1987). In addition only
chloroplasts donated by the PP cell react; BC-chloroplasts
remain unaffected. Eventually the BC-chloroplasts divide
and migrate to repopulate the hybrid cell with chloroplasts.

Not only is the incompatibility reaction found in the
hybrid cell itself but subsequently it also develops in
adjacent cells of the PP parent (e.g., cells SC1 & SC2, Fig.
2). The movement of the reaction away from the hybrid cell
can occur in either an acropetal or basipetal direction
depending on whether the PP-parent filament is attached
above or below the hybrid cell. The extent to which the
reaction moves is limited; if the PP-parent filament is long
enough, its more distant cells are not affected. This fact
and the kinetics of intercellular movement of the incompatibility response suggest that the reaction is not autocata-
lytic, and that after cell fusion an incompatibility-

generating substance is produced in the hybrid cell and
diffuses into adjacent cells (Koslowsky & Waaland, 1984).

In other systems, competition between organelles has
been invoked to explain uniparental inheritance of mitochondria and chloroplasts following sexual reproduction and
segregation of organelles following somatic cell fusion
(Birky, 1978; Bonnett & Glimelius, 1983; Eberhard,
1980,1981; Gressel et al., 1984). The mechanism by which
selective organelle destruction occurs is not understood.
The situation in Griffithsia is unusual in that incompatibility is expressed not only in the hybrid cell but also in
adjacent somatic cells that do not share a common cytoplasm.
A number of questions are raised by this incompatibility
response including: 1) what type of molecule is responsible
for selective chloroplast destruction; 2) why are only
chloroplasts affected; 3) what are the biochemical differences between the two types of chloroplasts; and 4) how does
the incompatibility-generating factor move from cell to
cell.   Incompatibility reactions have not been observed
following intraspecific cell fusions in other species of
Griffithsia or in Anotrichium tenue.

A second unusual result of somatic cell fusion has been
observed following fusion of cells from male and female
isolates of Anotrichium tenue (as G. tenuis; Waaland,
1978a).   Male/female hybrid cells were isolated and allowed
to regenerate new filaments.   In some instances, basipetal
cells of these new filaments formed reproductive structures
characteristic of the diploid generation (tetrasporangia);
spores produced by these tetrasporangia were sterile.   Cells
nearer the apex of the regenerating filament produced sexual
reproductive structures.   In Anotrichium, hybrid cells contain many nuclei (>100) from each parent; we do not know if
any nuclear fusion follows somatic cell fusion.   It is possible that male and female nuclei, when present in a single
cell, can cooperate to induce the production of tetrasporangia; one might predict that during subsequent cell
divisions the nuclei of one parent are segregated out so
that reproductive structures characteristic of the other
parent are then formed.   The genetic and biochemical control
of sexual morphogenesis in red algae is not well understood.
There are a number of reports of plants bearing either male
and female reproductive structures or tetrasporangia and
sexual structures (e.g., West & Norris, 1966, Rueness &
Rueness, 1973; Whittick & West, 1979). Inheritance of such

anomalies has been studied in Callithamnion tetragonium and Gracilaria tikvahiae (van der Meer & Todd, 1977; van der Meer et al. 1984; van der Meer, 1986; Rueness & Rueness, 1985). However, more research needs to be done to elucidate the cellular mechanisms that control reproductive morphogenesis: somatic hybrids may prove useful for such studies.

ANALYSIS OF CELL DIFFERENTIATION

At least two aspects of cell differentiation can be analyzed using the cell repair hormone, rhodomorphin: initiation of rhizoidal differentiation during regeneration and control of repair shoot cell differentiation. During cell regeneration from single shoot cells and shoot filaments, a shoot cell produces a different kind of cell, a rhizoid, at its base. One can ask, at what time during regeneration does the regenerating shoot cell base aquire rhizoid-specific characteristics. It has been shown that only rhizoids produce rhodomorphin (Waaland & Watson, 1980); therefore, the onset of rhodomorphin synthesis is a rhizoid-specific characteristic that can be used to measure rhizoid differentiation. Anotrichium has been used for these experiments because rhizoid regeneration by excised filaments occurs within 3.5-4 h and is nearly synchronous within a population of filaments. Rhodomorphin synthesis by individual rhizoids can be assayed using the following system: producer filaments with rhizoids are inserted rhizoid-first into cleaned Nitella wall cylinders and allowed to secrete rhodomorphin for 30 min. Producer filaments then are removed, and freshly decapitated test filaments are inserted apex first. If test filaments produce repair shoot cells, rhodomorphin is present. When excised filaments of Anotrichium are assayed for rhodomorphin production at 30 min intervals during regeneration, hormone secretion is detected beginning at 1.5-2 h after excision; this is about 2 h before cytokinesis will separate the new rhizoid from the shoot cell (Waaland, 1978b). Thus in Anotrichium cell differentiation begins well before cell division. It is of interest to know how this differentiation is controlled at a biochemical and molecular level. For example, are the nuclei at the base of a regenerating shoot cell directing a different developmental program than those at the top of the same cell. Because large numbers of cells at the same stage of development can be obtained in Anotrichium, it will be an excellent organism in which to explore questions of nuclear control of development in multinucleate cells.

Preliminary experiments have been conducted to explore the feasibility of studying repair shoot cell differentiation in G. pacifica (Waaland, 1986; Watson & Waaland, unpublished observations). Repair shoot cells resemble rhizoids in their morphology and in their pattern of cell elongation; however, the rhizoid-like growth and morphogenesis of repair shoot cells requires the presence of rhodomorphin. In the absence of rhodomorphin, repair shoot cells stop elongating and revert to a shoot cell pattern of morphogenesis (Waaland & Cleland, 1972; Waaland & Lucas, 1984). Repair shoot cells differ from both normal shoot cells and rhizoids in their insensitivity to unilateral light and in the fact that they are attracted to rhizoids. Prolonged treatment with rhodomorphin appears to shift the differentiation state of repair shoot cells. After 72-96 h in rhodomorphin, repair shoot cells lose their hormone dependence and continue to elongate even after rhodomorphin withdrawal; these cells now are morphologically and physiologically indistinguishable from rhizoids. Thus, repair shoot cells are delicately balanced in their differentiation between rhizoidal and shoot cell differentiation patterns and rhodomorphin appears to be able to shift that pattern. Since it is possible to use isolated rhodomorphin to obtain large quantities of repair shoot cells, it should be possible to characterize the molecular events that accompany shifts in cellular differentiation and morphogenesis.

CONTROL OF LOCALIZED GROWTH AND CYTOPLASMIC POLARITY

One developmental question that is important in all areas of biology, is how do cells develop and maintain polarity. Polarity is evident at the earliest stages of development (see Quatrano & Kropf, this volume). In differentiated cells, polarity is seen in localized secretion and localized growth. In Griffithsia, localized secretion and growth are seen in both rhizoid cells and repair shoot cells. These cells elongate by localized tip growth in which elongation is confined to the tip of apical cells (Waaland et al., 1972; Waaland & Lucas, 1984). Localized growth is reflected in a polar organization of cytoplasm at the tips of these cells. In the apical dome, there is a highly refractile zone of cytoplasm that contains nuclei, ribosomes, ER, and many active dictysomes, each with a mitochondrion on its forming face. Vesicles from the dictyosomes are seen to fuse with the plasma membrane in the apical dome of the cell (Waaland, 1983; 1987). Chloroplasts are excluded from the tip and form a deep red band just behind this refractile zone.

In order to evaluate subcellular factors that control elongation, it is useful if elongation can be manipulated by external factors just as the plane of polarity in Fucus can be altered using unilateral light (see Quatrano & Kropf, this volume). In rhizoids and repair shoot cells of Griffithsia, elongation can be changed by several factors. First we have seen that elongation of repair shoot cells can be turned off and on using the cell repair hormone rhodomorphin. In rhizoids, light controls elongation (Waaland, 1983; 1987). When rhizoids are transfered to continuous dark, they elongate for several hours. They then lose their refractile zones; at this time, organelles become redistributed and elongation ceases. When such cells are returned to the light, the refractile zone and chloroplast band are reorganized and elongation is reinitiated within 8 h (Waaland & Lucas, 1984). In addition rhizoids are negatively phototropic so the direction of their elongation can be shifted by exposure to unilateral light (Waaland et al., 1977).

We have used repair shoot cells and rhizoids of Griffithsia to test the hypothesis, first proposed by Jaffe and his co-workers, that transcellular currents are responsible for establishing and maintaining sites of localized secretion and growth (Jaffe, 1979; Weisenseel & Kircherer, 1981). Currents were measured using an extracellular vibrating probe microelectrode. Localized inflowing currents were detected at the tips of growing rhizoids and repair shoot cells (Waaland & Lucas, 1984). However, in repair shoot cells, the inflowing current continued even when rhodomorphin was withdrawn and elongation stopped; the current was not sufficient to maintain localized organelle accumulation and localized growth. In rhizoids that had ceased elongation in the dark, no current could be detected; however, current also could not be detected when such rhizoids were returned to the light and restarted growth. Thus in Griffithsia, transcellular currents per se do not appear to control localized organelle accumulation and localized growth. It is possible that gradients in specific ions may be important in these cells. Because organelle localization and localized growth can be easily manipulated in rhizoids and repair shoot cells, these cells are a good model for studying the role of such gradients and other factors that may be important in controlling localization and polarity.

Thus filamentous red algae are excellent organisms in which to study a variety of problems concerning the cellular control of developmental events. Studies have revealed the existence of a number of developmental processes that allow one to probe problems of morphogenesis including hormonal control of cell division, cell elongation and cell differentiation; and problems of organellar interaction. The control of reproductive morphogenesis, biochemical control of cell differentiation, and problems of subcellular polarity and localized secretion also may be investigated using filamentous red algal models.

ACKNOWLEDGEMENTS: This work was supported in part by N.S.F. Grant, DCB 8541028

REFERENCES

Birky CW (1978). The inheritance of genes in mitochondria and chloroplasts. BioScience 26:26-33
Bonnett HT, Glimelius K. (1983). Somatic hybridization in Nicotiana: behavior of organelles after fusion of protoplasts from male fertile and male sterile cultivars. Theor Appl Genet 65:213-217.
Dixon PS (1973). "Biology of the Rhodophyta." New York: Hafner Press.
Duffield ECS, Waaland SD, Cleland RE (1972). Morphogenesis in the red alga, Griffithsia pacifica: regeneration from single cells. Planta 105:185-195
Eberhard WG (1980). Evolutionary consequences of intracellular organelle competition. Q Rev Biol 55:231-249.
Eberhard WG (1981). Intraorganism competition involving eukaryotic organelles. Ann NY Acad Sci 361:44-52.
Gressel J, Cohen N, Binding H (1984). Somatic hybridization of atrazine resistant biotype of Solanum nigrum with Solanum tuberosus II. Segregation of plastomes. Theor Appl Genet 67:131-134.
Jaffe LF (1979). Control of development by ionic currents. In Cone RA, Dowling JE (eds): "Membrane Transduction Events," New York: Raven Press, pp 219-231.

Kochert G, Yates I (1974). Purification and partial characterization of a glycoprotein sexual inducer from Volvox carteri. Proc Natl Acad Sci USA 71:1211-1214.

Konrad-Hawkins E (1964). Developmental studies on regenerates of Callithamnion roseum Harvey. I. The development of a typical regenerate. Protoplasma 58:42-59.

Koslowsky DJ, Waaland SD (1984). Cytoplasmic incompatibility following somatic cell fusion in Griffithsia pacifica, a red alga. Protoplasma 123:8-17.

Koslowsky DJ, Waaland SD (1987). Ultrastructure of selective chloroplast destruction after somatic cell fusion in Griffithsia pacifica Kylin (Rhodophyta). J Phycol 23:638-648.

Lewis IF (1909). The life history of Griffithsia Bornetiana. Ann Bot (ns) 23:639-690.

L'Hardy-Halos M-Th (1971a). Recherches sur les Céramiacées (Rhodophycées, Céramiales) et leur morphogénèse. II. Les modalités de la croissance et les remaniements cellulaires. Rev Gén Bot 78:201-256.

L'Hardy-Halos M-Th (1971b). Recherches sur les Céramiacées (Rhodophycées, Céramiales) et leur morphogénèse. III. Observations et recherches expérimentales sur la polarit cellulaire et la hiérarchisation des éléments de la fronde. Rev Gén Bot 78:407-491.

L'Hardy-Halos M-Th, Larpent J-P (1983). Régénération chez les Algues. Rev Gén Bot 90:81-116.

Quatrano RS, Kropf DL (1989). Polarization in Fucus (Phaeophyceae) zygotes: investigations of role of calcium, microfilaments and cell wall. In Coleman AW, Goff LJ, Stein-Taylor JR (eds): "Alage as Experimental Syustems", New York: Alan R Liss, Inc, pp 111-119

Rueness J, Rueness M (1973). Life history and nuclear phases of Antithamnion tenuissimum with special reference to plants bearing both tetrasporangia and spermatia. Norw J Bot 20:205-210.

Rueness J, Rueness M (1985). Regular and irregular sequences in the life history of Callithamnion tetragonum (Rhodophyta, Ceramiales). Br Phycol J 20:329-333.

Schechter V (1934). Electrical control of rhizoid formation in the red alga, Griffithsia Bornetiana. J Gen Physiol 18:1-21.

Starr RC, Jaenicke L (1989). Cell differentiation in Volvox carteri Chlorophyceae): Use of mutants in understanding patterns and control. In Coleman AW, Goff LJ, Stein-Taylor JR (eds): "Alage as Experimental Syustems", New York: Alan R Liss, Inc, pp 135-147

Starr RC, Jaenicke L (1974). Purification and characterization of the hormone initiating sexual morphogenesis in
Volvox carteri f. nagariensis. Proc Natl Acad Sci USA
71:1025-1028.
van der Meer JP (1986). Genetics of Gracilaria tikvahiae
(Rhodophyceae). XI. Further characterization of a bisexual
mutant. J Phycol 22:151-158.
van der Meer JP, Todd ER (1977). Genetics of Gracilaria sp.
(Rhodophyceae, Gigartinales). IV. Mitotic recombination
and its relationship to mixed phases in the life history.
Can J Bot 55:2810-2817.
van der Meer JP, Patwary MU, Bird CJ (1984). Genetics of
Gracilaria tikvahiae (Rhodophyceae) X. Studies on a bisexual clone. J Phycol 20:42-46.
Waaland SD (1975). Evidence for a species-specific cell
fusion hormone in red algae. Protoplasma 86:253-261.
Waaland SD (1978a). Parasexually produced hybrids between
male and female plants of Griffithsia tenuis C. Agardh, a
red alga. Planta 138:65-68.
Waaland SD (1978b). Production of the cell fusion hormone,
rhodomorphin, by Griffithsia tenuis (Rhodophyta). J Phycol
14(Suppl):35.
Waaland SD (1981). In vitro cell fusion in the red alga
Griffithsia. Proc Internl Seaweed Symp 8:258-263.
Waaland SD (1983). Environmental control of rhizoid growth
and ultrastructure in Griffithsia pacifica (Rhodophyta,
Ceramiaceae). J Phycol 19(Suppl):6.
Waaland SD (1986). Hormonal coordination of the processes
leading to cell fusion in algae: a glycoprotein hormone
from red algae. In Bopp M (ed): "Plant Growth Substances
1985", Heidelberg: Springer-Verlag, pp 257-262.
Waaland SD (1987). Cell elongation in the red alga
Griffithsia: control by light, ions, and an endogenous
glycoprotein hormone. In Wiessner W, Robinson DG, Starr DC
(eds): "Algal Development. Molecular and Cellular
Aspects", Heidelberg: Springer-Verlag, pp 42-49.
Waaland SD, Cleland R (1972). Development in the red alga,
Griffithsia pacifica: control by internal and external
factors. Planta 105:196-204.
Waaland SD, Cleland RE (1974). Cell repair through cell
fusion in the red alga Griffithsia pacifica. Protoplasma
79:185-196.
Waaland SD, Lucas WJ (1984). An investigation of the role of
transcellular ion currents in morphogenesis of Griffithsia
pacifica Kylin. Protoplasma 123:184-191.

Waaland SD, Nehlsen W, Waaland JR (1977). Phototropism in the red alga, *Griffithsia pacifica*. Plant Cell Physiol 18:603-612.

Waaland SD, Waaland JR (1975). Analysis of cell elongation in red algae by fluorescent labelling. Planta (Berl) 126:127-138.

Waaland SD, Waaland JR, Cleland R (1972). A new pattern of plant cell elongation: bipolar band growth. J Cell Biol 54:184-190.

Waaland SD, Watson BA (1980). Isolation of a cell-fusion hormone from *Griffithsia pacifica* Kylin, a red alga. Planta 149:493-497.

Watson BA, Waaland SD (1983). Partial purification and characterization of a glycoprotein cell fusion hormone from *Griffithsia pacifica*, a red alga. Plant Physiol 71:327-332.

Watson BA, Waaland SD (1986). Further biochemical characterization of a cell fusion hormone from the red alga, *Griffithsia pacifica*. Plant Cell Physiol 27:1043-1050.

Weisenseel MH, Kicherer RM (1981). Ionic currents as control mechanisms in cytomorphogenesis. IN Kiermayer O (ed): "Cytomorphogenesis in Plants," Wien-New York: Springer, pp 379-399.

West JA, Norris RE (1966). Unusual phenomena in the life histories of Florideae in culture. J Phycol 2:54-57.

Westbrook MA (1927) *Callithamnion scopulorum* C.Ag. J Bot 65:129-138.

Whittick A, West JA (1979). The life history of a monoecious species of *Callithamnion* (Rhodophyta, Ceramiaceae) in culture. Phycologia 18:30-37.

Algae as Experimental Systems pages 135–147
© 1989 Alan R. Liss, Inc.

CELL DIFFERENTIATION IN <u>VOLVOX CARTERI</u> (CHLOROPHYCEAE):
THE USE OF MUTANTS IN UNDERSTANDING PATTERNS AND CONTROL

Richard C. Starr and Lothar Jaenicke

Department of Botany, The University of Texas
at Austin, Austin, Texas 78713 (R.C.S.) and
Institut für Biochemie, An der Bottmühle 2,
D-5000 Köln 1, FRG (L.J.)

For the investigator who would study the control of cell
differentiation, the green algal genus <u>Volvox</u> provides a
model of ultimate simplicity with its two cell types, yet
exhibits a desirable degree of complexity in the variations
that may occur in these two cell types. Eighty years have
passed since Powers (1907,1908) noted the potential of <u>Volvox</u>
as a research tool for studies in development, and 24 years
ago Barth (1964) pointed out that the family of colonial
flagellates to which <u>Volvox</u> belongs presents "the problem of
the fundamental control of somatic versus germ cell differentiation in such challenging simplicity that one wonders why
contemporary investigators do not turn to them as materials
for fresh inroads into the problem."

The potential for using <u>Volvox</u> in research on cell
differentiation was greatly increased with the pioneering
study on sexual induction in <u>Volvox aureus</u> by Darden (1966)
which was followed closely by similar studies on other
species; but for the most part there has been little positive
response from those developmental biologists who are seeking
to understand how different cell types in a multicellular
organism develop during the successive cleavages of a single
cell to form a multicellular individual. <u>Drosophila</u>, <u>Caeno-
rhabditis</u>, <u>Arabidopsis</u>, and other complex multicellular animals and plants still dominate the thinking and the research
efforts of developmental biologists. It may not be the
"developmental biologist" who will solve the problem of cell
differentiation but rather some new breed of cellular and
molecular biologist not bound by the traditional thought
patterns of the classical or neo-classical developmentalist.

This discussion is directed to that investigator with a new idea and a new technique who is looking for a unique experimental system.

The development of Volvox as an organism for experimentation was made possible as a result of the formulation of a defined medium in which the alga could be cultured axenically (Provasoli & Pintner, 1959). Darden's (1966) successful use of this medium for V. aureus was followed by studies on a variety of species by other investigators (V. carteri f. weismannia, Kochert, 1968; V. carteri f. nagariensis, Starr, 1969; V. rousseletii, McCracken & Starr, 1970; V. pocockiae and V. spermatosphaera, Starr, 1970b; V. gigas and V. powersii, Vande Berg & Starr, 1971; V. africanus, Starr, 1971; V. dissipatrix, Starr, 1972; V. obversus, Karn et al., 1974; V. capensis, Miller & Starr, 1981). Of these, one species has proved to have the combination of characteristics that has made it the "Volvox of choice" by current investigators: Volvox carteri f. nagariensis Iyengar, a dioecious heterothallic species isolated from a rice paddy near Kobe, Japan, in 1967 (Starr, 1969, 1970a).

The genus Volvox is characterized by having 1000 or more cells arranged on the perphery of a spheroid, but in any spheroid there are only two types of cells: small, biflagellated somatic cells and larger reproductive cells. The reproductive cells may be gonidia (asexual reproductive cells), eggs, or sperm-producing cells. In some species both eggs and sperm-producing cells may be in the same spheroid (monoecious); in others they occur only in separate spheroids (dioecious). In dioecious species, sexual reproduction may occur within a clonal population (homothallism) or the two sexes may be produced in separate clonal populations (heterothallism). In those species where there are no distinct female spheroids with eggs, the gonidia of the asexual spheroid function as facultative eggs. Table I shows the variation among the different species.

The asexual spheroids in both the female (HK10) and the male (69-1b) strains of V. carteri f. nagariensis are identical, having approximately 2000 small biflagellated somatic cells and a maximum of 16 gonidia (asexual reproductive cells) (Fig. 1). Unlike species of Volvox in other Sections of the genus, V. carteri and other species in the Section Merrillosphaera have no cytoplasmic connections between the cells in the mature spheroid, although they do have them in the embryo.

TABLE I

Variation of Sexual Expression Among Species of <u>Volvox</u>

	HOMOTHALLIC	HETEROTHALLIC	MONOECIOUS	DIOECIOUS	FEMALES WITH EGGS	FACULTATIVE FEMALES
V. carteri f. weismannia		X		X	X	
V. carteri f. nagariensis		X		X	X	
V. gigas		X		X	X	
V. obversus		X		X	X	
V. powersii	X			X	X	
V. spermatosphaera	X				X	X
V. africanus						
(U.S.)	X			X	X	
(Australia)		X		X	X	
(S. Africa)	X		X			
(India)	X		X+males			
V. tertius	X				X	X
V. aureus	X				X	X
V. pocockiae	X				X	X
V. dissipatrix						
(Australia)	X		X			
(India)		X		X	X	
V. barberi	X		X			
V. capensis	X		X			
V. globator	X		X			
V. rousseletii		X		X	X	

Female spheroids resemble asexual spheroids in having approximately 2000 somatic cells, but there are 35 or more smaller, darker eggs instead of the 16 gonidia (Fig. 2). Male spheroids are dwarfs with no more that 512 cells, often 128 or 256, in the spheroid; there is a 1:1 ratio of small somatic cells and the larger sperm-producing cells (Figs. 3, 4).

The most exciting aspect of <u>Volvox</u> <u>carteri</u> <u>f.</u> <u>nagariensis</u> as a material for the study of cell differentiation is the fact that new spheroids are formed by a series of successive cleavages of the gonidia; the asexual, the female and the male spheroids show a particular pattern of cleavage in which

the respective reproductive cells are delimited by readily
observable unequal cleavages of cells at predictable times
and places in the embryogenesis of each spheroid.

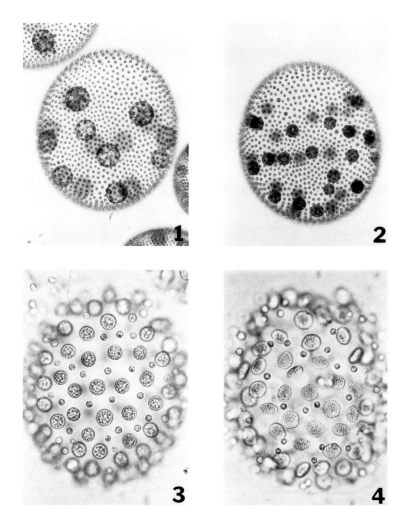

Figs. 1-4. Volvox carteri f. nagariensis
    Fig. 1. Asexual spheroid with 16 gonidia. X100
    Fig. 2. Female spheroid with 35+ eggs. X100
    Fig. 3. Male spheroid - young. X250
    Fig. 4. Male spheroid - with sperm packets. X250

In the development of the asexual embryo unequal cleavage of certain cells at the division of the 32-celled stage produces the 16 gonidial initials. Even when there are fewer than 16 gonidia formed, the pattern of formation has been shown to be predictable (Gilles & Jaenicke, 1982). In embryogenesis of the female the 35+ eggs are formed by unequal divisions of certain cells in the 64-celled embryo, whereas in the male, the 1:1 ratio of sperm-producing cells to somatic cells is a direct consequence of the unequal cleavage of every cell at the last division in embryogenesis.

The geometrical pattern of cell cleavages in the complete embryogenesis of Volvox was first described by Janet (1923) in a beautiful study of V. aureus. This proved to be an invaluable guide in elucidating the cell lineages during the embryogenesis of V. carteri. The formation of the asexual embryo in V. carteri has been the subject of a number of detailed investigations and subsequent reviews, therefore there is little reason to discuss this in further detail here (Kochert, 1968, 1975; Starr, 1969, 1970a; Viamontes & Kirk, 1977; Viamontes et al., 1979; Green & Kirk, 1981, 1982; Gilles & Jaenicke, 1982; Jaenicke & Gilles, 1982; Kirk et al., 1982; Kirk & Harper, 1986).

Not only can V. carteri f. nagariensis be grown asexually under controlled conditions in an axenic defined medium, but the sexual forms can be elicited easily by the addition of an inducer formed by the sperm. This inducer was partially purified by Starr and Jaenicke (1974), and the nature of the molecule and its biochemical action continues to occupy researchers in several laboratories. Its protein sequence including the attachment sites of the N-glycosylating oligosaccharides has been elucidated from the appropriate cDNA by Mages, Tschochner, and Sumper (July, 1988) at Lehrstuhl Biochemie I, Universität Regensburg, FRG, and R. Gilles (June, 1988, personal communication) at the Institut für Biochemie, Universität Köln, FRG. The inducer glycoprotein is active at the very low concentration of 6 X 10^{-17}M. Both male and female strains of V. carteri respond to the inducer such that the pattern of cleavage in embryogenesis is changed, resulting in the formation of male spheroids and female spheroids in the respective strains. Although certain biochemical events can be detected as a result of the addition of the inducer, the relation of these events to any internal change in the gonidium which will now form a sexual embryo is as yet unknown (Gilles et al., 1984; Wenzl & Sumper, 1986)

Until recently the species-specific inducer was the only means by which the pattern of embryogenesis could be changed, but now we know of other ways to accomplish induction. This has been investigated using a mutant strain obtained by UV irradiation of the wild-type female strain (HK10). The mutant was first identified by its formation of females with at least twice the number of eggs seen in a wild-type female (Fig. 5). A more interesting trait was a 10-fold increase in sensitivity to the inducer. The two characteristics (egg number, increased sensitivity) have not been seen to segregate when the strain is crossed with a wild-type male, nor are they linked to sex. From these crosses a strain (designated as Gone 12) was selected for study because it was more robust that the original mutant isolate.

Several treatments of young spheroids of Gone 12 have resulted in the formation of sexual spheroids by their gonidia without the use of inducer. Gone 12 is routinely grown for experimentation at 28C, under cool white fluorescent light of 12,000 lux on a 16h/8h, light/dark cycle, resulting in a 48h generation time. For the following treatments young spheroids at the beginning of the light peroid on the second day were used:

1) <u>Ultraviolet irradiation</u>. In the course of development of a protocol for use of UV irradiation as a mutagenic agent, it was observed that long exposures (6-7 min, 500uW/sec) would often result in the formation of sexual spheroids by gonidia which had not been killed. After some experimentation it was found that many gonidia of young spheroids of Gone 12 irradiated for only 2 minutes at 500 uW/sec and immediately returned to the illuminated shelves would form sexual embryos if the gonidia were removed from the spheroids by drawing the spheroids into a Pasteur pipette with a bore smaller than the diameter of the spheroid. If irradiated spheroids were left intact, the gonidia would form only asexual embryos.

2) <u>Aldehydes</u>. Using anthranilic acid formalide as an agent to kill somatic cells of a young spheroid resulted in the production of sexual embryos by the gonidia (Starr & Jaenicke, 1988).

3) <u>Lectins</u>. Young spheroids were placed in Volvox medium (pH 6.75) containing 10mg/ml of a variety of lectins (Con A, ECA, HPA, PNA, SJA, STA, WGA, and PHA from Sigma Chemical Co., St. Louis). With the exception of PHA (from <u>Phaseolus vulgaris</u>)

none had any effect. As the young spheroids began to enlarge in the PHA solution, the layer of somatic cells on the periphery became disrupted until eventually the gonidia were left enclosed only in the non-cellular internal matrix. Embryogenesis resulted in the formation of female spheroids. Various fractions of the PHA obtained from Sigma did not all cause induction. Highly purified PHA-P and PHA-E were effective, but PHA-L and PHA-M were not; PHA-P conjugated with FITC or with biotin was not effective.

4) <u>Protease XIV</u> (<u>Sigma</u>). This highly non-specific protease at a concentration of 100 ug/ml caused the dissolution of young spheroids within 45 min. The gonidia were washed with gentle centrifugation in several changes of medium and placed in Volvox medium (pH 6.75) in small Petri dishes. Subsequent cleavage of the treated gonidia resulted in the formation of female embryos. It is interesting that earlier reports of protease treatments failed to result in sexual embryos in the wild-type strain (HK10) even when using the inducer glycoprotein at a concentration known to induce the

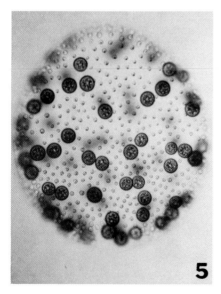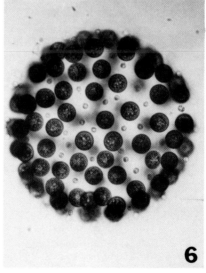

Figs. 5, 6. <u>Volvox</u> <u>carteri</u> <u>f</u>. <u>nagariensis</u>
    Fig. 5. Gone 12 - multiegg female. X150
    Fig. 6. Mutant male with eggs. X250

gonidia in intact spheroids (Gilles & Jaenicke, 1982). Later
it was shown that induction was possible with released gonidia
of the HK10 strain provided the concentration of the inducer
was increased 100 times that needed to induce gonidia in
intact spheroids (Wenzl & Sumper, 1986). In the present study
it was possible to induce gonidia of the HK10 strain without
added inducer, but only if they were removed by the protease
within the first hour after the onset of illumination; in the
more sensitive Gone 12, the protease treatment was effective
if done within the first 3-4 h after the onset of illumination.
The lack of success by others may well have been due to this
narrow window during which time the gonidia are susceptible
to self-induction.

   Although, under normal conditions, only the sperm secrete
inducer into the medium, Kirk and Kirk (1986) have shown that
under heat stress the female strain becomes self-inducible
due to secretion of inducer in such large quantities that it
can be assayed in the surrounding medium. In light of the
results reported above in which successful induction of the
female was obtained through alternative methods not involving
exogenous inducer, we propose that the gonidia of the female
strain are secreting inducer even under standard growth conditions, but in such small quantities that neither is it detectable in the surrounding medium nor does it cause self-
induction. It is possible that self-induction is prevented
by the binding of this small amount of inducer to some component secreted by the somatic cells into the matrix. Thus
it is only when the somatic cells are removed from the young
spheroid (or killed by aldehydes) and the inhibiting factor
already present is inactivated by UV irradiation, aldehydes,
or proteases, that the isolated gonidium becomes self-induced
by its own secretion of inducer. This is also suggested since
induction of the gonidia isolated with protease does not occur
in the presence of serum containing antibodies against the
purified inducer.

   Although the alternate protocols for induction were developed using the more sensitive Gone 12, the wild-type strain
HK10 also responded positively to all treatments except the
PHA lectin. A critical factor in all treatments was the time
at which the treatment occurred. Unlike Gone 12, which was
reactive for the first 3-4 h after the onset of illumination,
the HK10 strain was reactive only during the first hour of
illumination. The PHA lectin effectively removes the somatic
cell layer of both Gone 12 and HK10, but the time required

(2-4h) exceeds the narrow window of time during which HK10 is responsive to any alternative treatment.

Exogenously supplied inducer in the induction process appears to saturate the neutralizing factor in the matrix and may not necessarily be involved directly in the induction process. This could well account for the extremely small amounts of inducer needed for the induction of the wild-type HK10 and the 10-fold increase in sensitivity noted in Gone 12. This increased sensitivity could be due to either a greater production of the endogenous inducer by the gonidia or a decrease in the production of the neutralizing factor of the matrix. When homogenates of young asexual spheroids of Gone 12 (obtained by sonication within 2-3h after the onset of illumination) are bioassayed with Gone 12, induction occurs in tubes containing 1% of the sonicate, again indicating the presence of inducer in the untreated spheroids.

A variety of mutations have been described in V. carteri f. nagariensis, both spontaneous (Starr, 1970a) and induced with physical and chemical agents (Huskey et al., 1979). We have been using UV irradiation to induce mutations affecting the development of the sexual spheroids in the male and female strains. Irradiation of the female strain (HK10) followed by the addition of inducer yielded the sensitive multiegg strain from which Gone 12 was derived. Irradiation of the male strain followed by induction produced two different mutants which showed defects in the reproductive cells that normally would have undergone cleavage to form sperm packets. In both mutants the spheroids showed the typical male pattern of a 1:1 ratio of somatic to reproductive cells. In one mutant the reproductive cells developed as asexual reproductive cells (gonidia); a mutant of this type was already described by Callahan and Huskey (1980). The second mutant of the male strain formed spheroids with the typical male morphology, but after enlargement the reproductive cells darkened in a manner similar to that observed in egg cells of the female spheroids (Fig. 6). When sperm packets from the wild-type male strain (Pal 3), from which the mutant had been derived, were added to these mutant spheroids, the sperm packets penetrated the spheroids and in 4-5 days typical orange zygotes could be seen. When the zygotes germinated, the progeny showed a 1:1 ratio of normal sperm-producing males and mutant "egg-producing males."

Inasmuch as Volvox, like other green flagellates, is

haploid throughout its life cycle except for the zygote phase,
one must conclude that in monoecious species with both sperm
and eggs in the same spheroid, or in homothallic dioecious
species where separate male and female spheroids are formed
in the same clonal population, the genetic pathways for both
sexes are present in the same haploid genome. The same must
be true even for the heterothallic dioecious species such as
V. carteri f. nagariensis in which male and female spheroids
are formed in separate clonal populations, when one can
demonstrate, as described above, that a sexually competent
egg-bearing spheroid strain may result from the mutagenic
action on the male member of a heterothallic pair of strains.
This is the first instance in which one sex of a heterothallic
pair of strains in Volvox, or any other green alga, has been
changed by mutation to the other sex.

Sexual expression in V. carteri f. nagariensis depends
on a sequence of genes under the control of a locus which is
activated either directly or indirectly by the glycoprotein
inducer. In some manner this control locus in the male activates only those genes involved in the pattern of male
spheroid differentiation, followed by the activation of those
genes involved in the production of sperm by the reproductive
cells of the male; an allelic locus functions as the activator
of the female sequence in the other of the heterothallic pair
of strains. The egg-bearing mutant male is the result of
some genetic change which is effective only after embryogenesis and serves to point out the genetic complexity of
the sexual process.

We now know much more about V. carteri f. nagariensis
that we did at the time of its isolation 20 years ago (Starr,
1969); however, as exciting as some of the results have been,
it is obvious that we have only defined the problems more
clearly rather than finding the answers to the questions about
the mechanism of cell differentiation. When one reads the
excellent exhaustive review by Kirk and Harper (1986) entitled
"Genetic, Biochemical and Molecular Approaches to Volvox
Development and Evolution" one cannot help but notice how
often, following a discussion of experimental data, it is
necessary to include in the closing paragraphs such statements
as "No model yet proposed appears to adequately explain all
the observed facts of gonidial specification" or "The only
thing that is abundantly clear by now is that the more this
locus is studied, the more interesting its behavior seems to
be" or "Nevertheless, the concept that this may be the

mechanism by which the sexual inducer acts in Volvox is a
novel departure and certainly warrants further attention."
That similar conclusions will not summarize future reviews
of cell differentiation in Volvox can only be hoped.

REFERENCES

Barth LJ (1964). "Development: Selected Topics." Reading,
    MA: Addison-Wesley.
Callahan AM, Huskey RJ (1980). Genetic control of sexual
    development in Volvox. Dev Biol 80:419-435.
Darden WH Jr (1966). Sexual differentiation in Volvox aureus.
    J Protozool 13:239-255.
Gilles R, Gilles C, Jaenicke L (1984). Pheromone-binding and
    matrix-mediated events in sexual induction of Volvox carteri.
    Z Naturforsch 39c:584-592.
Gilles R, Jaenicke L (1982). Differentiation in Volvox car-
    teri: study of pattern variation of reproductive cells.
    Z Naturforsch 37c:1023-1030.
Green KJ, Kirk DL (1981). Cleavage patterns, cell lineages,
    and development of a cytoplasmic bridge system in Volvox
    embryos. Cell Biol 91:743-755.
Green KJ, Kirk DL (1982). A revision of the cell lineages
    recently reported for Volvox carteri embryos. J. Cell Biol
    94:741-742.
Huskey RJ, Griffin BE, Cecil PO, Callahan AM (1979). A pre-
    liminary genetic investigation of Volvox carteri. Genetics
    91:229-244.
Jaenicke L, Gilles R (1982). Differentiation and embryogene-
    sis in Volvox carteri. In Jaenicke L (ed): "Biochemistry
    of Differentiation and Morphogenesis," Berlin:Springer-
    Verlag, pp 288-294.
Janet C (1923). "Le Volvox. Troisieme Memoire." Macon
    Protat Freres.
Karn RC, Starr RC, Hudock GA (1974). Sexual and asexual
    differentiation in Volvox obversus (Shaw) Printz, strains
    Wd3 and Wd7. Arch Protistenk 116:142-148.
Kirk DL, Harper JF (1986). Genetic, biochemical, and mo-
    lecular approaches to Volvox development and evolution.
    Int Rev Cytol 99:217-293.
Kirk DL, Kirk MM (1986). Heat shock elicits production of
    sexual inducer in Volvox. Science 231:51-54.
Kirk DL, Viamontes GI, Green KJ, Bryant JL Jr (1982). Inte-
    grated morphogenetic behavior of cell sheets: Volvox as a
    model. In Subtelny S, Green PB (eds): "Developmental Order:

Its Origin and Regulation," New York: Alan R. Liss, pp 247-274.

Kochert G (1968). Differentiation of reproductive cells in *Volvox carteri*. J Protozool 15:438-452.

Kochert G (1975). Developmental mechanisms in *Volvox* reproduction. In Markert CL, Papaconstantinou J (eds): "The Developmental Biology of Reproduction," New York: Academic Press, pp 55-90.

Mages H-W, Tschochner H, Sumper M (1988). The sexual inducer of *Volvox carteri*, primary structure deduced from cDNA sequence. FEBS Letters 234(2):407-410.

McCracken MD, Starr RC (1970). Induction and development of reproductive cells in *Volvox rousseletii*. Arch Protistenk 112:262-282.

Miller CM, Starr RC (1981). The control of sexual morphogenesis in *Volvox capensis*. Ber Deutsch Bot Ges 74:357-372.

Powers JH (1907). New forms of *Volvox*. Trans Am Microsc Soc 27:123-149.

Powers JH (1908). Further studies in *Volvox*, with descriptions of three new species. Trans Am Microsc Soc 28 141-175.

Provasoli L, Pintner IJ (1959). Artificial media for freshwater algae; problems and suggestions. In Tryon CA, Hartman RT (eds): "The Ecology of Algae," Spec Publ #2, Pymatuning Lab. of Field Biology: University of Pittsburgh, pp 84-96.

Starr RC (1969). Structure, reproduction, and differentiation in *Volvox carteri f. nagariensis*, strains HK 9 & 10. Arch Protistenk 111:204-222.

Starr RC (1970a). Control of differentiation in *Volvox*. Dev Biol Suppl 4:59-100.

Starr RC (1970b). *Volvox pocockiae*, a new species with dwarf males. J Phycol 6:234-239.

Starr RC (1971). Sexual reproduction in *Volvox africanus*. In Parker BC, Brown RM (eds): "Contributions in Phycology," Lawrence, KA: Allen Press, pp 59-66.

Starr RC (1972). Sexual reproduction in *Volvox dissipatrix*. Brit Phycol J 7:284 (abst)

Starr RC, Jaenicke L (1974). Purification and characterization of the hormone initiating sexual morphogenesis in *Volvox carteri f. nagariensis*. Proc Nat Acad Sci (USA) 71:1050-1054.

Starr RC, Jaenicke L (1988). Sexual induction in *Volvox carteri f. nagariensis* by aldehydes. Sex Plant Reprod 1:28-31.

Vande Berg WJ, Starr RC (1971). Structure, reproduction, and differentiation in *Volvox gigas* and *Volvox powersii*. Arch

Protistenk 113:195-219.

Viamontes GI, Fochtmann LJ, Kirk DL (1979). Morphogenesis in
  Volvox: analysis of critical variables. Cell 17:537-550.

Viamontes GI, Kirk DL (1977). Cell shape changes and the
  mechanism of inversion in Volvox. J Cell Biol 75:719-730.

Wenzl A, Sumper M (1986). Early event of sexual induction in
  Volvox: chemical modification of the extracellular matrix.
  Dev Biol 115:119-128.

Algae as Experimental Systems pages 149–167
© 1989 Alan R. Liss, Inc.

CELLULAR MORPHOGENESIS: THE DESMID (CHLOROPHYCEAE) SYSTEM

Oswald Kiermayer and Ursula Meindl

Institute of Plant Physiology, University of
Salzburg, A-5020 Salzburg, Austria

Cytomorphogenesis is one of the most mysterious phenomena in biology. Although it has been possible to elucidate
important cellular morphogenesis processes in recent years
the basic mechanisms leading to special forms and patterns
are still obscure. Included in the entire process of cellular morphogenesis is not only the formation of external cell
shape and ornamentation of the cell wall but also the formation of the intracellular architecture e.g. the internal
shaping, localization and anchoring of various organelles
such as the nucleus, the chloroplast and others. Thus, scientific work on problems of cytomorphogenesis must encompass
the full range of formative events taking place during cell
differentiation.

Unicellular desmids are among the most favourable organisms for experimental studies of cytomorphogenesis. These
green algae, especially the genera Micrasterias, Cosmarium,
Closterium, Pleurenterium and some others, are especially
suitable for this kind of investigation because of their
symmetric cell pattern and because they are easily cultiva-
ted under artificial culture conditions. With these cells it
is possible to follow the full developmental sequence and
morphogenetic events within a short range of time during
which a great number of different cytological, cytochemical,
ultrastructural or experimental investigations can be undertaken. The distinct structural differences between the primary wall and the secondary wall in many desmids make a separate study of the underlying mechanisms of their growth
and patterning possible. In addition, desmids are characterized by clearly marked ultrastructural elements such as the

golgi-complex, the cytoskeleton with different microtubule systems and a number of characteristic vesicles. Experimental studies on the inner cell architecture, established by special migrations of the chloroplast and the nucleus, which had been found to be under the control of cytoskeletal elements offer new insights into the mechanism of organelle motion and orientation and their relation to cytomorphogenesis.

Seen from a historical point of view, Pringsheim (1918), was the first to cultivate desmids under artificial conditions. In 1950 Waris introduced a culture medium ("Waris solution") especially for growing <u>Micrasterias</u>. The use of this artificial culture medium, together with elegant experimental light microscopic techniques extended by Kallio (1951), made experimental studies on desmids, especially on the influence of the nucleus on cytomorphogenesis, possible for the first time. In 1961 Drawert and Mix started the first investigations on ultrastructure of desmids. After a series of osmotic studies on <u>Micrasterias</u> cells (Kiermayer & Jarosch, 1962; Kiermayer, 1964) the latter author introduced a special fixation procedure for electron microscopy. This yielded satisfactory results clarifying the ultrastructure of differentiating <u>Micrasterias</u> cells (Kiermayer, 1968a), especially the roles of microtubules, the golgi-system, various vesicles etc. Reviews of investigations on cytomorphogenesis in desmids, especially on <u>Micrasterias</u>, are given by Brook (1981), Kallio and Lethonen (1981), Kiermayer (1981), Pickett-Heaps (1983) and Kiermayer and Meindl (1984).

METHODICAL APPROACHES TO CYTOMORPHOGENESIS IN DESMIDS

A considerable number of different techniques had been used in studies of desmid morphogenesis.

1) <u>Culture methods</u>: Semisterile cultures of various desmids originating from natural habits are obtained by a washing process under aseptic conditions. After some transfers the washed cells are inoculated into Erlenmeyer flasks containing a defined sterilized growth medium (e.g. "Waris medium"; Waris, 1950) and placed into an illuminated incubator. Developmental stages used in most of the described experiments on cytomorphogenesis can be obtained by special regulation of the light-dark cycles (see Kiermayer, 1980).

2) Light microscopy and cinemicrography: Cell pairs of
a distinct developmental stage of various desmids are transferred from the culture flasks into small glass dishes where
they can be experimentally treated with different drugs.
Cells of each sample are examined in the light microscope.
For studies of growth processes or organelle movements time-
lapse cinemicrography (16 mm film) is useful (Kiermayer,
1965; Meindl, 1987a). For quantitative frame by frame analysis a Kodak "Analyst" film projector can be used (see
Meindl, 1986).

3) Osmotic studies: Interesting effects on primary wall
and pattern formation can be obtained by turgor reduction.
For this purpose developmental stages are placed into solutions of polyethylene glycol, glucose or mannitol and examined
with the light microscope (Kiermayer, 1964; Kobayashi, 1973;
Tippit & Pickett-Heaps, 1974; Ueda & Yoshioka, 1976).

4) Cytochemistry: An indirect immunofluorescence method
using mouse monoclonal antibodies against alpha- and beta-
tubulin of chicken brain can be used for identification of
microtubules in desmids (Hogetsu & Oshima, 1985). For detection of membrane-associated Ca^{2+}, developmental stages are
treated with chlorotetracycline (CTC) and are examined using
an incident light fluorescence microscope (Meindl, 1982a,b).
Autoradiography with (H^3)methylmethionine in which the label
is incorporated by methylation into wall polysaccharides is
described by Lacalli (1975).

5) Physical treatment: Local UV- or laser-irradiation
may be used to inactivate the nucleus or parts of the cytoplasm (Kallio & Heikkilä, 1969; Lacalli & Acton, 1972). Centrifugation can be employed for the dislocation of organelles
(Kallio 1951; Kiermayer, 1981). Application of electric fields
to Micrasterias developmental stages produce effects on morphogenesis and membranes (Brower & McIntosh, 1980). Ionic
currents can be studied with the vibrating probe technique
in desmids (Troxell et al., 1986).

6) Electron microscopy: For ultrastructural studies of
desmid cell developmental stages, fixation is performed with
1% glutaraldehyde for 10 min ("short time fixation"; Kiermayer, 1968a) followed by postfixation with OsO_4, dehydration
and embedding into epon (see Kiermayer, 1968a; Meindl, 1983).
For freeze-etching, desmid cells are prefixed in 1% glutaraldehyde, infiltrated with 30% glycerol, rapidly frozen in

Freon and freeze-etched on a Balzers freeze-etch apparatus
(Staehelin & Kiermayer, 1970; Kiermayer & Sleytr, 1979). A
method using a propane-jet device for rapid freezing gives
excellent results (Giddings et al., 1980). For scanning
electron microscopy the algae are fixed with glutaraldehyde
and OsO_4 and dried in a "critical point-drying" device. The
dried specimens are coated with carbon and gold and are
examined in a scanning electron microscope (Pickett-Heaps,
1973; Neuhaus & Kiermayer, 1982).

   7) Theoretical models: Recently, theoretical models
for studying morphogenesis in Micrasterias were introduced
by Lacalli & Harrison (1987) using Turing´s model for
branching tip growth. Another new approach to cytomorphogenesis is application of theories of formal languages and
reproducing automata combined with the knowledge about developmental sequences (Pohja & Lethonen, 1985).

EXPERIMENTAL STUDIES ON CYTOMORPHOGENESIS

Cell Shape Formation

   Cell development in desmids studied so far starts during
mitosis with the formation of a septum which grows inward
centripetally separating the two half cells from each other.
Subsequently the primary wall is formed, and each semicell
regenerates a new half. After a few hours, primary wall growth
is finished and the secondary wall with its regulary arranged cell wall pores is deposited on its inner face. Finally,
the primary wall is shed and the secondary wall remains as
the only wall layer. Deviations from this course of cell
development, which is true for most of the studied desmids,
have been described in cells like Closterium which do not
shed their primary wall but possess a three layered cell
wall with pores only in its outer sheath (Brook, 1981).

   Ultrastructural studies revealed that septum formation
in Micrasterias, Cosmarium and Pleurenterium occurs almost
in the same way, beginning with the formation of a girdle of
wall material located between the two separating secondary
walls from which the three layered septum wall grows (Kiermayer, 1981; Meindl, 1987b). Numerous golgi vesicles, the
"septum vesicles" (SEV), which vary in size and form among
the different genera (Meindl, 1987b) are found in the area

of the growing septum (Fig. 1). They are assumed to be functional in septum formation by fusion with the plasma membrane.
The contents, which occur as dark portions within the septum
material probably establish the septum wall while their
membranes contribute material for the growing plasma membrane.

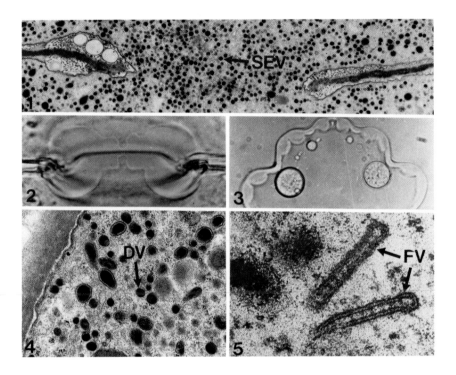

Figure 1. Septum vesicles (SEV) in Micrasterias denticulata
involved in septum formation; x 8,000. Fig. 2,3, "Accumulation pattern" of wall material induced by turgor reduction
in M. denticulata. Fig. 2, "Initial pattern" of septum;
x 1,300. Fig. 3, Pattern of older developmental stage; x 830.
Fig. 4,5, Various vesicle types in M. denticulata. Fig. 4,
Dark vesicles (DV) involved in primary wall formation;
x 21,000. Fig. 5, Flat vesicles (FV) involved in secondary
wall formation; x 128,000 (Kiermayer, 1981).

Studies with turgor reduction in Micrasterias reveal

that at the stage of septum formation a pre-pattern of the
later cell shape is present. This "septum initial pattern"
(Kiermayer, 1970b, 1981), which becomes visible by a patterned deposition of wall material during turgor reduction, is
characterized by two lateral and one central minimum zone
(Fig. 2) which correspond to the lobe invaginations formed
during the later developmental sequence. These experiments
show that a characteristic species-specific pre-pattern can
already be detected at the septum stage long before actual
shape formation starts (Kiermayer, 1970b).

Primary wall growth in desmids occurs in different ways.
In Micrasterias it can be regarded as multipolar tip growth
where, after septum formation, each semicell develops a
bulge which forms sequentially symmetric invaginations leading to the characteristic cell shape (Kiermayer, 1964; Lacalli, 1975). In Cosmarium and Closterium the unipolar tips
of the young half cells round off already at a quite early
stage of cell development (Pickett-Heaps & Fowke, 1970;
Pickett-Heaps, 1972), whereas the initial growth in Pleurenterium is characterized by an expansion of the lateral areas
of the cell wall during which the apex of the growing half
cells remains flattened (Meindl, 1987b).

The primary wall of the desmids investigated thus far
is composed of a network of loosely aggregated microfibrils
(Kiermayer, 1970a). In freeze-etch studies of Micrasterias
and Closterium it has been demonstrated that during synthesis
of the primary wall the plasma membrane carries single "rosettes" whereas during secondary wall formation hexagonally
arranged "rosettes" are obviously involved in microfibril
formation (Kiermayer & Sleytr, 1979; Giddings et al., 1980;
Hogetsu, 1983). A relationship between microtubules and "rosettes" had been observed by Giddings and Staehelin (1988).
Different types of vesicles are produced by the dictyosomes
during primary wall growth in desmids (Kiermayer, 1970a; Noguchi, 1978). They may be subdivided into "large vesicles"
(LV), probably containing mucilage and "dark vesicles" (DV)
probably involved in primary wall formation (Fig. 4; Kiermayer, 1970a; Pickett-Heaps & Fowke, 1970; Meindl, 1987b).

Experiments with turgor reduction in osmotic solutions
during the different stages of primary wall formation demonstrated (Kiermayer, 1964; Ueda & Yoshioka, 1976) that under
these conditions the primary wall material, which in untreated cells is used for cell wall extension and shaping, is

deposited in a stage-specific "accumulation pattern" while cell extension ceases (Fig. 3). These experiments clearly show that wall material is excreted at local areas of the cell periphery, probably by fusion of dark vesicles (Kiermayer, 1970a,b).

In consequence of the ultrastructural studies and the experiments with turgor reduced cells, especially in Micrasterias it had been assumed that the plasma membrane of growing half cells bears patterned "membrane recognition sites" for membranes of the primary wall material delivering dark vesicles. By fusion of these vesicles with the "recognition sites" of the plasma membrane corresponding to a species-specific template, cell wall growth and thus lobe formation may occur at distinct predetermined areas during the developmental sequence (Kiermayer, 1970b, 1981). A chemical characterization of the postulated "membrane recognition sites" in Micrasterias had been obtained by using chlorotetracycline (CTC), a fluorescent indicator for membrane-associated Ca^{2+} (Meindl, 1982a). The experiments with CTC revealed that during primary wall formation each developmental stage of Micrasterias is characterized by patterned distributed fluorescent areas at the cell periphery (Fig. 6-9). Since these areas correspond to the actual growth zones of the cell it seems likely that the "membrane recognition sites" postulated by Kiermayer are areas of accumulated membrane-associated Ca^{2+} from which it is supposed vesicle fusion is promoted. A correlation between the cell pattern and the distribution of the Ca^{2+} accumulation areas was also demonstrated in studies on the defect mutant cell Micrasterias thomasiana f. uniradiata (Meindl, 1985a) which lacks cell pattern at one side of the cell (Waris, 1950). Corresponding to the asymmetric distribution of wall material accumulation areas formed during turgor reduction, the mutant exhibits fluorescent zones after CTC treatment only at that side of the half cell which forms a normal cell pattern during the developmental sequence (Fig. 9; Meindl, 1985a), and shows no fluorescent areas at the "defect" side. Since studies with the ionophore A23187 (McNally et al., 1983; Lethonen, 1984) are in good agreement with these results, the involvment of membrane-associated Ca^{2+} in cell shaping of Micrasterias seems to be proved. Recently, studies with the vibrating probe technique on ion fluxes in Micrasterias and Closterium have shown that current patterns are correlated with developing wall patterns. The currents seem to be carried in part by Ca^{2+} (Troxell et al., 1986). The requirement of Ca^{2+} for

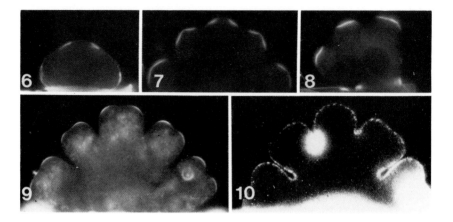

Figures 6-9. Patterned arrangement of fluorescent spots in cortical area of different developmental stages of Micrasterias cell during growth of primary wall treated with chlorotetracycline, indicating accumulation of membrane-associated Ca^{2+}. Fig. 6,7,9, M. denticulata; x 600 (Meindl, 1982a). Fig. 8, Mutant cell M. thomasiana f. uniradiata lacking cell pattern at one side; x 800 (Meindl, 1985a). Fig. 10, Patterned arrangement of fluorescent dots in cortical area of M. denticulata cell, treated with chlorotetracycline during formation of cell wall pores; x 600 (Meindl, 1982b).

vesicle fusion at the cell surface in various cell systems has been recently discussed by Steer (1988). Whereas in Micrasterias the plasma membrane seems to control patterned cell wall growth and thus morphogenesis via a membrane recognition process, in Closterium additionally microtubules are obviously involved in cell shaping (Pickett-Heaps & Fowke, 1970). A disturbance of these microtubules by colchicine leads to severe changes of cell shape and to alterations in the microfibril orientation of the cell wall (Hogetsu & Shibaoka, 1978) whereas in Micrasterias microtubule inhibitors do not influence shape formation (Kiermayer, 1981). The significance of actin microfilaments for cytomorphogenesis in Micrasterias is discussed by Ueda and Noguchi (1988).

When primary wall formation of the desmid cell is complete the secondary wall, which differs in structure and

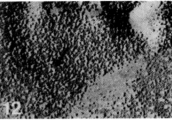

Figure 11,12. Micrasterias denticulata. Fig. 11, Bundles of
microfibrils of secondary wall after freeze-etching;
x 50,000 (Kiermayer & Sleytr, 1979). Fig. 12, "Rosettes"
(arrow) in the plasma membrane of growing cell after freeze-
etching, probably involved in formation and orientation of
microfibrils of secondary wall; x 107,000 (Kiermayer &
Sleytr, 1979).

formation from the primary wall, is deposited on its inner
face. The secondary wall is composed of parallel oriented
microfibrils which aggregate laterally to form layers of
crossed bands (Fig. 11; Kiermayer, 1981). Kiermayer and
Dobberstein (1973) found that during secondary wall formation in Micrasterias a particular type of disk-like vesicle,
the "flat vesicles" (FV) are delivered from the dictyosomes.
These FV carry sack-like structures at their edges and are
characterized by an unusual thick membrane that contains
special globular particles of about 20 nm at their inner
membrane surface arranged in an hexagonal pattern (Fig. 5).
By fusion with the plasma membrane these globular particles
reach the outside of the plasma membrane where they are incorporated. Freeze-etch studies of the plasma membrane during
the stage of secondary wall formation have demonstrated
(Kiermayer & Sleytr, 1979; Giddings et al., 1980) that the
globular particles represent little rosettes each composed
of 6 particles (Fig. 12). Since the same periodicity can be
found in the lattice of the rosettes of the plasma membrane
and the bands of microfibrils of the cell wall, the rosettes
seem to be responsible for the formation and orientation of
microfibrils (Kiermayer & Sleytr, 1979; Giddings et al.,
1980; see Hogetsu, 1983 for Closterium). The distribution of
the rosettes of the plasma membrane may be perturbed by
applied electric fields (Brower & McIntosh, 1980). A similar
formation of the secondary wall by the action of F-vesicles

PRIMARY WALL GROWTH
(Fusion of D-Vesicles)

PORE FORMATION
(Fusion of P-Vesicles)

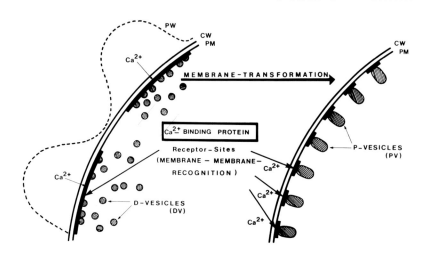

Figure 13. Hypothetical representation of possible interaction processes between golgi products and plasma membrane during primary wall growth and pore formation: CW (left side)=primary wall during bulge stage; CW (right side)=primary wall at the end of its growth (higher magnification); PW=primary wall during lobe formation; PM=plasma membrane (Kiermayer & Meindl, 1984).

had also been found in various other species of Micrasterias (Noguchi, 1978; Meindl, 1985a) and in Pleurenterium tumidum (Staurastrum tumidum; Meindl, 1987b). In Penium spirostriolatum the secondary wall which is structurally similar to that of Micrasterias is formed in another way (Mix & Manshard, 1977). Here secondary wall formation is obviously mediated by special elongated cytoplasmic vesicles that contain an amorphous matrix with a bundle of 8-12 fibrils. These fibrils probably are used to build up the microfibril bands of the secondary wall.

The secondary wall of desmids is penetrated by pores which are formed during secondary wall growth and are distributed in a distinct species specific pattern, (e.g., Neuhaus & Kiermayer, 1982). Their formation is mediated by special "pore vesicles" (PV) that probably fuse with distinct spots of the plasma membrane to form a plug in the growing secondary wall. The incorporation of the "pore vesicles" into the plasma membrane at distinct areas produces a specific pore pattern, which is probably determined by a "membrane re-cognition process" similar to that postulated for patterned primary wall growth.(Kiermayer, 1981). A pattern of fluores-cent dots corresponding to the pore pattern can be observed on the cell periphery at the beginning of secondary wall for-mation in Micrasterias in CTC-treated cells (Fig. 10; Meindl, 1982b). The dots indicate "fusion sites" for pore vesicles. These results suggest that membrane-associated Ca^{2+} may not only be critical for control of local vesicle fusion during tip growth but may also promote fusion of pore vesicles with the plasma membrane at discrete sites (Fig. 13; Meindl, 1982b; Kiermayer & Meindl, 1984). The various mentioned ve-sicles (e.g., SEV, DV, FV, possibly PV) are all derivatives of the dictyosomes produced sequentially during cell diffe-rentiation. It seems that a temporal "switching" of the dic-tyosomes leading to a sequential production of different stage-specific types of vesicles is operating during the whole process of cellular morphogenesis (Kiermayer, 1981; Kiermayer & Meindl, 1984). Unfortunately, nothing is known about this "switching process" itself.

Formation and Maintenance of Intracellular Architecture

The intracellular architecture of the desmid cell is primarily determined by the arrangement of the chloroplast and the location of the nucleus. In all constricted desmids studied so far a cytoskeleton-mediated immigration of the chloroplast into the growing half cell takes place (Kiermayer, 1968b). In these cells the chloroplast of the two half cells separates when cell development is finished. In unconstricted desmids such as Netrium or Closterium the cleavage of the chloroplast takes place in both half cells and the chloro-plast division products each then occupy half of each off-spring cell.(Carter, 1919, 1920; Pickett-Heaps & Fowke, 1970).

Nuclear migration is also an important event for the establishment of the inner cell architecture. Three types of nuclear migration during cell development of desmids have been reported. 1) In the constricted genera Micrasterias and Cosmarium the nucleus moves into the growing semicell during cell development and returns to its central position when

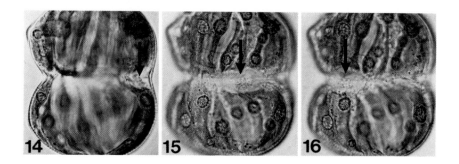

Figures 14-16. Circular nuclear migration in Pleurenterium tumidum; Serial photographs taken at 15 min intervals; Note change of nuclear position (arrow); x 550 (Meindl, 1986).

cell differentiation is finished (Kiermayer, 1964; Pickett-Heaps, 1972). 2) In unconstricted cells like Closterium, Netrium, Hyalotheka or Cylindrocystis, the nuclei migrate into each non-growing semicell along a groove in the chloroplast and are displaced into the cell center by expansion of the whole cell content (Carter, 1919, 1920; Pickett-Heaps & Fowke, 1970). Both types of nuclear migration are mediated by cytoskeletal elements. During its migration the nucleus is ensheathed by a microtubule "cage" that arises from a microtubule center and as demonstrated for Micrasterias also contains microfilaments (Fig. 17; Kiermayer, 1968a; Pickett-Heaps & Fowke, 1970; Pickett-Heaps, 1972; Meindl, 1983, 1984, 1985b). Anti-microtubule drugs strongly influence nuclear migration by disturbance of this microtubule system and lead to a dislocation of the nucleus (Kiermayer, 1968b; Meindl & Kiermayer, 1981). Anchorage of the nucleus after its migration and during interphase is probably caused by a ring-shaped microtubule band running along the cell periphery of the isthmus area in all desmids studied so far ("isthmus band of microtubules"; Kiermayer, 1968a). 3) In Pleurenterium tumidum an extraordinary nuclear migration which had not been previously described in other organisms was recently detected (Meindl, 1986). At the end of primary wall formation the nucleus moves radially towards the cortical cytoplasm of the isthmus area, penetrates the chloroplast lobes and performs autonomous circular migrations lasting up to three days along the cell wall of the isthmus (Fig. 14-16; Meindl, 1986, 1987a). Ultrastructural studies suggest that the nucleus moves on a hoop-like

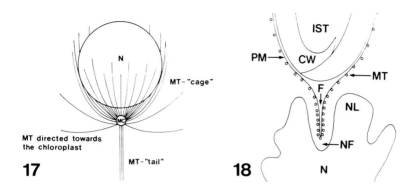

Figures 17, 18. Schematic drawings of microtubule arrangement during nuclear migration in desmids. Fig. 17, Micrasterias denticulata: N=nucleus; MC=microtubule center; MT=
microtubules (Meindl, 1983). Fig. 18, Nuclear area during
circular nuclear migration in Pleurenterium tumidum seen
from a median view: CW=two overlapping cell walls of the
two half cells; F=fold of plasma membrane; IST=isthmus area;
NF=nuclear furrow; NL=nuclear lobe; PM=plasma membrane
(Meindl, 1986).

track formed by a fold of the plasma membrane and an adjacent
band of microtubules which the nucleus surrounds in a deep
furrow of its cell wall-facing surface (Fig. 18; Meindl,
1986). Circular nuclear migration is inhibited by the anti-
microtubule drug amiprophos-methyl (APM), whereas cytochalasin B does not stop the movement, although cytoplasmic streaming is completely blocked (Meindl, 1986, 1987a). When circular nuclear migration is finished the nucleus moves back
to the center of the isthmus.

Nuclear Control of Cytomorphogenesis

The influence of the nucleus on cytomorphogenesis in
Micrasterias was studied by producing anucleate, binucleate
and diploid cells especially by centrifugation during mitosis (Kallio, 1951). One of the most important results of
these studies is the observation of a quantitative relationship between nuclear and cytoplasmic mass. An increase of
nuclear quantity (e.g., in binucleate or polyploid cells) is

correlated with an increase in the size and the complexity
of the new half cell, whereas a decrease of the nuclear mass
in anucleate cells leads to a reduction of the cell pattern
(for references see Kallio & Lethonen, 1981). An elimination
of the nuclear control in Micrasterias can also be obtained
by UV-irradiation (Kallio & Heikķilä, 1969) or by treatment
with various inhibitors of RNA or protein synthesis (Selman,
1966; Tippit & Pickett-Heaps, 1974; Noguchi & Ueda, 1979;
Kiermayer, 1981). The effects caused by these physical or
chemical treatments may be summarized as "anuclear type of
development (ATD)" (Selman, 1966) syndrome. This ATD-syndrome,
which after cycloheximide treatment is observed in biradiate
Micrasterias cells, in the mutant M. thomasiana f. uniradiata
(Meindl, 1985a), in Xanthidium armatum and in Closterium
lunula (Kiermayer & Meindl, 1986), is characterized among
others by swelling of the young semicells, decrease of growth
velocity, total inhibition of secondary wall formation and
a severe disturbance of the ultrastructure of the primary
wall, probably due to an inhibition of the normal fusion of
the primary wall material containing D-vesicles (for references see Kiermayer 1981; Kiermayer & Meindl, 1986). These results indicate that a continuous nuclear control is essential
for normal cell development and cell shaping in desmids. Experimental studies on genetic mechanisms would greatly enhance
our knowledge on cytomorphogenesis but unfortunately, thus
far, desmids have proved to be difficult objects for this
kind of investigations because of their reduced sexuality.

SUMMARY

This article has shown that desmid cells are used in a
great variety of experimental studies to elucidate basic
processes in cell biology, especially in cellular morphogenesis. With different physiological, ultrastructural and cytochemical methods it has been demonstrated that the plasma
membrane primarily controls patterned growth of the primary
wall and pore formation probably via "recognition processes"
mediated by local Ca^{2+} accumulation and vesicle fusion. In
some cases e.g., Closterium, cell shaping may also be influenced by cytoskeletal elements. A pre-pattern for cytomorphogenesis exists already at the septum stage. The pattern
of aggregation of microfibrils in the secondary wall corresponds to the pattern of hexagonal arrays of "rosettes" which
are incorporated into the plasma membrane by the fusion of
dictyosomal membrane templates (F-vesicles). For the produc-

tion of different vesicle types a sequential "switching" of
the dictyosomes during the course of cytodifferentiation obviously takes place. Continuous structural and functional
transformation of the plasma membrane by a sequential incorporation of the membranes of these vesicles seems to be a
basic mechanism for cytomorphogenesis. Various experiments
clearly demonstrate that a continuous nuclear control and
protein synthesis is essential for cellular morphogenesis in
desmids. For the establishment and maintenance of the inner
cell architecture, especially, various microtubule systems
in association with microfilaments are functional controlling
chloroplast and nuclear migration and nuclear anchoring.
Also, for further research work focused on the basic processes described in this articel the desmid system represents
an ideal experimental object.

REFERENCES

Brook AJ (1981). "The Biology of Desmids". Oxford: Blackwell
    Scientific Publ, pp 185-195.
Brower DL, McIntosh JR (1980). The effect of applied electric
    fields on Micrasterias. I. Morphogenesis and the pattern
    of cell wall deposition. J Cell Sci 42:261-277.
Carter N (1919). Studies on the chloroplasts of desmids. I.
    Ann Bot 33:215-254.
Carter N (1920). Studies on the chloroplasts of desmids. IV.
    Ann Bot 34:305-319.
Drawert H, Mix M (1961). Licht- und elektronenmikroskopische
    Untersuchungen an Desmidiaceen. II. Mitt. Hüllgallerte und
    Schleimbildung bei Micrasterias, Pleurotaenium und Hyalo-
    theka. Planta 56:237-261.
Giddings TH, Brower DL, Staehelin LA (1980). Visualization
    of particle complexes in the plasma membrane of Micraste-
    rias denticulata associated with the formation of cellulose
    fibrils in primary and secondary walls. J Cell Biol 84:327-
    339.
Giddings TH, Staehelin LA (1988). Spatial relationship be-
    tween microtubules and plasma membrane rosettes during de-
    position of primary wall microfibrils in Closterium.
    Planta 173:22-30.
Hogetsu T (1983). Distribution and local activity of particle
    complexes synthesizing cellulose microfibrils in the plasma
    membrane of Closterium acerosum (Schrank) Ehrenberg. Plant
    Cell Physiol 24:777-781.
Hogetsu T, Oshima Y (1985). Immunofluorescence microscopy of

microtubule arrangement in Closterium acerosum (Schrank) Ehrenberg. Planta 166:169-175.

Hogetsu T, Shibaoka H (1978). Effects of colchicine on cell shape and on microfibril arrangement in the cell wall of Closterium acerosum. Planta 140:15-18.

Kallio P (1951). The significance of nuclear quantity in the genus Micrasterias. Ann Bot Soc Zool Bot Fenn Vanamo 24:1-222.

Kallio P, Heikkilä H (1969). UV-induced facies change in Micrasterias torreyi. Österr Bot Z 116:226-243.

Kallio P, Lethonen J (1981). Nuclear control of morphogenesis in Micrasterias. In Kiermayer O (ed): "Cytomorphogenesis in Plants", Cell Biol Monographs Vol. 8 Wien New York: Springer, pp 191-213.

Kiermayer O (1964). Untersuchungen über die Morphogenese und Zellwandbildung bei Micrasterias denticulata Brêb. Protoplasma 59:382-420.

Kiermayer O (1965). Micrasterias denticulata (Desmidiaceae)-Morphogenese. Film E 868, Inst Wiss Film Göttingen.

Kiermayer O (1968a). The distribution of microtubules in differentiating cells of Micrasterias denticulata Brêb. Planta 83:223-236.

Kiermayer O (1968b). Hemmung der Kern- und Chloroplastenmigration von Micrasterias durch Colchizin. Naturwissenschaften 55:299-300.

Kiermayer O (1970a). Elektronenmikroskopische Untersuchungen zum Problem der Cytomorphogenese von Micrasterias denticulata Brêb. I. Allgemeiner Überblick. Protoplasma 69:97-132.

Kiermayer O (1970b). Causal aspects of cytomorphogenesis in Micrasterias. Ann NY Acad Sci 175:686-701.

Kiermayer O (1980). Control of morphogenesis in Micrasterias. In Gantt E (ed): "Handbook of Phycological Methods: Developmental and Cytological Methods", Cambridge: Cambridge University Press, pp 6-12.

Kiermayer O (1981). Cytoplasmic basis of morphogenesis in Micrasterias. In Kiermayer O (ed): "Cytomorphogenesis in Plants". Cell Biol Monographs Vol. 8 Wien New York: Springer, pp 147-189.

Kiermayer O, Dobberstein B (1973). Membrankomplexe dictyosomaler Herkunft als "Matrizen" für die extraplasmatische Synthese und Orientierung von Mikrofibrillen. Protoplasma 77:437-451.

Kiermayer O, Jarosch R (1962). Die Formbildung von Micrasterias rotata Ralfs. und ihre experimentelle Beeinflussung. Protoplasma 54:382-420.

Kiermayer O, Meindl U (1984). Interaction of the golgi-apparatus and the plasmalemma in the cytomorphogenesis in Micrasterias. In Wiessner W, Robinson D, Starr RC (eds): "Compartments in Algal Cells and Their Interaction". Berlin, Heidelberg: Springer, pp 175-182.

Kiermayer O, Meindl U (1986). Das "Anuclear type of development" (ATD)-Phänomen, hervorgerufen durch Hemmung der Proteinsynthese, bei verschiedenen Desmidiaceen. Ber Nat-Med Ver Salzburg 8:101-114.

Kiermayer O, Sleytr U (1979). Hexagonally ordered "rosettes" of particles in the plasma membrane of Micrasterias denticulata Brêb. and their significance for microfibril formation and orientation. Protoplasma 101:133-138.

Kobayashi S (1973). Relationship between cell growth and turgor pressure in Micrasterias. Bot Mag (Tokyo) 86:309-313.

Lacalli TC (1975). Morphogenesis in Micrasterias. I. Tip growth. J Embryol Exp Morph 33:95-116.

Lacalli TC, Acton AB (1972). An inexpensive laser microbeam. Trans Amer Micros Soc 91:236-238.

Lacalli TC, Harrison LG (1987). Turing's model and branching tip growth: relation of time and spatial scales in morphogenesis with application to Micrasterias. Can J Bot 65:1308-1319.

Lethonen J (1984). The significance of Ca^{2+} in morphogenesis of Micrasterias studied with EGTA, Verapamil, $CaCl_3$ and calcium ionophore A 23187. Plant Sci Lett 33:53-60.

McNally JG, Cowan JD, Swift H (1983). Effect of the ionophore A 23187 on pattern formation in the alga Micrasterias. Develop Biol 97:137-145.

Meindl U (1982a). Local accumulation of membrane-associated calcium according to cell pattern formation in Micrasterias denticulata, visualized by chlorotetracycline fluorescence. Protoplasma 110:143-146.

Meindl U (1982b). Patterned distribution of membrane-associated Ca^{2+} during pore formation in Micrasterias. Protoplasma 112:138-141.

Meindl U (1983). Cytoskeletal control of nuclear migration and anchoring in developing cells of Micrasterias denticulata and the change caused by the anti-microtubular herbicide amiprophos-methyl (APM). Protoplasma 118:75-90.

Meindl U (1984). Helical structures in the cytoplasm of differentiating cells of Micrasterias thomasiana. Protoplasma 123:230-232.

Meindl U (1985a). Experimental and ultrastructural studies on cell shape formation in the defect mutant cell Micrasterias thomasiana f. uniradiata. Protoplasma 129:74-87.

Meindl U (1985b). Aberrant nuclear migration and microtubule arrangement in a defect mutant cell of Micrasterias thomasiana. Protoplasma 126:74–90.

Meindl U (1986). Autonomous circular and radial motions of the nucleus in Pleurenterium tumidum and their relation to cytoskeletal elements and the plasma membrane. Protoplasma 135:50–66.

Meindl U (1987a). Pleurenterium tumidum (Desmidiaceae): Circular motions of the nucleus and their relation to cytoskeletal elements. Film E 3018 Encyclopaedia Cinematographica, Göttingen.

Meindl U (1987b). Zellentwicklung und Ultrastruktur der Desmidiacee Pleurenterium tumidum. Nova Hedwigia 45:347–373.

Meindl U, Kiermayer O (1981). Biologischer Test zur Bestimmung der Antimikrotubuli-Wirkung verschiedener Stoffe mit Hilfe der Grünalge Micrasterias denticulata. Mikroskopie (Wien) 38:325–336.

Mix M, Manshard E (1977). Über Mikrofibrillen-Aggregate in langgestreckten Vesikeln und ihre Bedeutung für die Zellwand bei einem Stamm von Penium (Desmidiales). Ber Deutsch Bot Ges 90:517–526.

Neuhaus G, Kiermayer O (1982). Raster-elektronenmikroskopische Untersuchungen an Desmidiaceen: Die Poren und ihr Verteilungsmuster. Nova Hedwigia 36:499–568.

Noguchi T (1978). Transformation of the golgi-apparatus in the cell cycle, especially at the resting and earliest developmental stages of a green alga Micrasterias americana. Protoplasma 95:73–88.

Noguchi T, Ueda K (1979). Effect of cycloheximide on the ultrastructure of cytoplasm in cells of a green alga Micrasterias crux-melitensis. Biol Cell 35:103–110.

Pickett-Heaps JD (1972). Cell division in Cosmarium botrytis. J Phyc 8:343–360.

Pickett-Heaps JD (1973). Stereo-scanning electron microscopy of desmids. J Microsc (Oxford) 99:109–116.

Pickett-Heaps JD (1983). Morphogenesis in desmids: Our present stage of ignorance. Modern Cell Biol 2:241–258.

Pickett-Heaps JD, Fowke LC (1970). Mitosis, cytokinesis and cell elongation in the desmid Closterium littorale. J Phycol 6:189–215.

Pohja P, Lethonen J (1985). Simulation of morphogenesis and speciation in the unicellular alga Micrasterias (Chlorophyta, Conjugatophyceae). J Theor Biol 115:401–414.

Pringsheim EG (1918). Die Kultur der Desmidiaceen. Ber Deutsch Bot Ges 36:482–485.

Selman GG (1966). Experimental evidence for the nuclear control of differentiation in Micrasterias. J Embryol Exp Morph 16:469-485.

Staehelin LA, Kiermayer O (1970). Membrane differentiation in the golgi-complex of Micrasterias denticulata Brêb. visualized by freeze-etching. J Cell Sci 7:787-792.

Steer MW (1988). The role of calcium in exocytosis and endocytosis in plant cells. Physiologia Plant 72:213-220.

Tippit DH, Pickett-Heaps JD (1974). Experimental investigations into morphogenesis in Micrasterias. Protoplasma 81: 271-296.

Troxell CT, Scheffey C, Pickett-Heaps JD (1986). Ionic currents during wall morphogenesis in Micrasterias and Closterium. In Nuccitelli R (ed): "Ionic Currents in Development". New York: AR Liss Inc, pp 105-112.

Ueda K, Noguchi T (1988). Microfilament bundles of F-actin and cytomorphogenesis in the green alga Micrasterias cruxmelitensis. Europ J Cell Biol 46:61-67.

Ueda K, Yoshioka S (1976). Cell wall development of Micrasterias americana, especially in isotonic and hypertonic solutions. J Cell Sci 21:617-631.

Waris H (1950). Cytophysiological studies on Micrasterias. I. Nuclear and cell division. Physiologia Plant 3:1-16.

Section III.
Cellular Recognition

Algae as Experimental Systems pages 171–185
© 1989 Alan R. Liss, Inc.

RECOGNITION PROTEINS OF CHLAMYDOMONAS REINHARDTII (CHLOROPHYCEAE)

Ursula W. Goodenough and W. Steven Adair

Biology Department, Washington University, St. Louis, Missouri 63130 (U.W.G.), and Department of Anatomy and Cell Biology, Tufts Medical School, Boston, Massachusetts 02111 (W.S.A.)

Chlamydomonas reinhardtii, a unicellular haploid alga, responds to nitrogen starvation by leaving the vegetative (mitotic) phase of its life cycle and differentiating into a gamete. Gametes are of two complementary mating types (plus and minus) and interact in a highly selective manner. The simplicity of this inducible cell-cell recognition system, together with the facility with which Chlamydomonas can be manipulated genetically (Levine and Ebersold, 1960), has attracted the attention of numerous investigators over the years.

In 1977, we initiated a collaboration to study this system, and have worked together for 11 years. During this time we have written several review articles (Adair, 1985, 1987, 1988; Goodenough 1977, 1985; Goodenough and Adair, 1980; Goodenough and Ferris, 1987; Goodenough and Thorner, 1983; Goodenough et al, 1980, 1986a). Rather than write yet another formal review, we have elected to describe the sequence of experimental approaches and insights that occurred during our collaboration. While somewhat unorthodox, we hope that such an account will prove useful to those embarking on the use of algae as experimental systems. It is important to stress at the outset that several other groups were involved in similar research during this same period. While our personal account does not, by definition, encompass their work, it is well described in several reviews (see van den Ende et al, this volume; Snell, 1985; van den Ende, 1985; Musgrave and van den Ende, 1987).

IDENTIFICATION OF THE AGGLUTININS

By 1977, the C. reinhardtii mating system was well understood in broad outline. The two mating types were known to adhere to one another by a set of complementary factors, displayed on their flagellar surfaces, which Wiese (1969) designated "agglutinins". Wiese (1965) had further shown that if gametes of one mating type are allowed to swim in distilled water for 24 h the supernatant came to contain "gamones" which, when presented to gametes of the opposite mating type, caused them to isoagglutinate. Gametes not only adhered by their flagella in a manner reminiscent of mating, but also shed their cell walls, a normal prelude to zygotic fusion. Based on these observations, Wiese suggested that "gamones" are equivalent to agglutinins. In the mid-1970s, however, Bergman et al (1975) and Snell (1976) demonstrated that "gamones" consist of vesiculated flagellar membrane vesicles, not agglutinin molecules per se. This accounted for the multivalent nature of the material and led several workers to examine vesicle components for agglutinin candidates.

Analysis of vesicle preparations by sodium dodecyl sulfate polyacrylamide gel electrophoresis (SDSPAGE) yielded little information. Polypeptide profiles were dominated by the major flagellar membrane glycoprotein (350 kD), and no differences were detected between vesicles isolated from plus and minus gametes (Bergman et al, 1975). Moreover, no reproducible differences were observed between gametic vesicles and material isolated from vegetative (non-agglutinating) cells and agglutination-defective (imp) mutants (Bergman et al, 1975). These observations led us to suspect that the 350 kD flagellar membrane glycoprotein might function as a common adhesin, acquiring its sex-specific properties (e.g., novel glycosylations) during gametogenesis.

To test whether there were diagnostic immunological differences between the plus and minus membranes, rabbit polyclonal antibodies were raised against plus and minus membrane preparations. The resulting antisera readily isoagglutinated gametes, with the agglutinating foci moving to the flagellar tips ("tipping"), in the fashion of sexual adhesion (Goodenough and Jurivich, 1978). However, both antisera elicited these responses from both plus and minus gametes, and reacted strongly with the flagellar surfaces of vegetative cells and imp mutants as well. To analyze the specificity of antisera more closely, therefore, we devised

the gel overlay procedure (Adair et al, 1978), a precursor
to Western blotting. Flagellar preparations, resolved by
SDSPAGE, were incubated with anti-plus or anti-minus
antisera, followed by ^{125}I protein A from the bacterium S.
aureus, a reagent that detects IgG proteins.
Autoradiography revealed that each antiserum recognized
multiple bands in addition to the ubiquitous 350 kD. Again,
no mating-type or gamete-specific species were apparent.
Moreover, all attempts to adsorb the antisera against one
mating type to obtain a sex-specific probe were failures, as
were attempts to recover agglutinin activity from 350 kD
material isolated from gels.

At this point we learned of a new non-ionic detergent,
octylglucoside (OG), that could be easily removed by
dialysis. We therefore extracted gametic flagella with OG,
removed the detergent by dialysis, and assayed the material
for agglutinin activity. Initial observations were
disappointing: neither mating type-specific
isoagglutination nor a block in adhesiveness were found.
Closer inspection, however, revealed that over a period of a
few minutes, "tester" gametes, while not isoagglutinated,
adhered strongly to the microscope slide. When a dialyzed OG
extract was applied directly to the slide and allowed to
dry, gametes displayed a rapid, strong agglutination
response. Moreover, flagellar adhesion was mating type-
specific and elicited the normal mating responses (flagellar
tipping, wall loss). Finally, extracts prepared from
vegetative cells and imp mutants were inactive. This led to
the development of a simple quantitative bioassay (Adair et
al, 1982), setting the stage for identification and
isolation of agglutinin molecules.

As a prelude to agglutinin identification, methods were
developed for selectively surface labeling gametes with
^{125}I. Autoradiography of surface-labeled flagellar
preparations and OG extracts revealed a complex pattern
which included multiple flagellar-associated species as well
as contaminating wall components (Monk et al, 1983). As
expected, the profile was dominated by 350 kD. While no
obvious agglutinin candidates were evident in the starting
material, we had the necessary tools (quantitative bioassay
and sensitive detection system) to begin the hunt.

At this point, serendipity stepped in. During this
period, we were also studying the role of external divalent
cations on gametic cell fusion. In one experiment, plus
gametes incubated in EDTA displayed no flagellar adhesion

when mixed with untreated minus gametes. However, many
gametes adhered to the slide by their flagellar tips,
reminiscent of the response to OG extracts. Similar to OG
extracts, EDTA extracts proved to be mating type-specific
and elicited the expected biological responses from "tester"
gametes (Adair et al, 1982). These findings established
that active agglutinin could be extracted from living cells
without detergents, implying that the molecules were
extrinsic. EDTA extracts, which contained a much higher
titer of agglutinin activity and were much simpler to
prepare, quickly became the preferred source of starting
material for agglutinin purification.

SDSPAGE/autoradiography of EDTA extracts from ^{125}I-
surface labeled plus gametes contained little if any 350 kD,
thereby putting to rest its candidacy as the agglutinin.
There were, however, numerous other bands, including cell
wall glycopolypeptides that invariably contaminated our
flagellar preparations and EDTA extracts. While a modest
enrichment was achieved by ammonium sulfate precipitation,
the composition remained complex. Preliminary attempts to
fractionate agglutinin activity by ion-exchange and
hydrophobic interaction chromatography were unsatisfactory,
and led us to try gel filtration. Agglutinin activity was
found to be excluded from a Sepharose 6B matrix, implying a
very high molecular weight and/or a very large Stokes
radius, a finding subsequently confirmed by sucrose gradient
centrifugation which showed that agglutinin sediments as a
12S species (Adair et al, 1982).

It is important at this point to review the scientific
context of our experiments. In early 1981, the only large
extrinsic protein characterized to any extent was
fibronectin, and its diverse biological activities were only
beginning to be appreciated. Our existing biases,
therefore, had programmed us to expect a molecule like
Concanavalin A, the then-prototypic "surface recognition
molecule," which was hardly a 12S species. We suspected,
therefore, that the apparent large size of agglutinin was
due to aggregation and devoted much attention to optimizing
the composition of column buffers in an attempt to disrupt
"aggregates". Agglutinin continued to behave as an
extremely large species, although we did note that the
presence of OG significantly improved separation from the
majority of contaminating polypeptides. With Sepharose 6B
chromatography, therefore, significant purification was
achieved.

Looking back on our publication from that period (Adair et al, 1982), the agglutinin leaps out at us: the only SDSPAGE band conspicuously unique to the void-volume fraction in our Sepharose 6B column migrates in the 3% stacking gel of Laemmli gels (Adair et al, 1982, Figs. 8,9). However, polypeptides were not expected to be retained in stackers, and such material was routinely attributed to non-specific aggregation. Indeed, in early experiments this portion of the gel was often removed prior to drying.

In the spring of 1982, things began to come together. Agglutinin activity in the void volume of the Sepharose 6B matrix was fractionated to homogeneity on the newly available large-pore matrix Fractogel-75, and we finally realized that the purified plus agglutinin corresponded to the material migrating in the stacking gel. We then devised a method to put our non-agglutinating (imp) mutants to work for us. Mutants were surface-labeled, EDTA-extracted, and the extracts co-chromatographed with unlabeled EDTA extracts from wild-type cells. This allowed us to assay column fractions for biological activity (wild type-derived), and analyze corresponding labeled (mutant-derived) material by SDSPAGE/autoradiography. In all cases, mutant extracts were found to lack the stacking-gel species co-fractionating with agglutinin activity (Adair et al, 1983a). The final proof of identity came from analyzing complementing and non-complementing diploid cells, constructed from imp mutants by PEG fusion. Complementing, but not non-complementing, diploids displayed agglutinating activity and possessed a molecule comigrating with wild-type agglutinin. Thus we had achieved convincing biochemical and genetic evidence for agglutinin identity. But what kind of molecule would fail to move through a 3% stacking gel?

FORMULATING THE HYPOTHESIS

Here we can give credit to a very important component of productive science: one's scientific environment. Shortly after the identity of the plus agglutinin was established, three collaborators joined us from other laboratories at Washington University to make key technical and intellectual contributions. The first was John Heuser, who had recently applied the quick-freeze deep-etch technique to visualizing proteins adsorbed to mica flakes (Heuser, 1983). When he examined the peak fraction from the Fractogel-75 column by this procedure, he encountered a startling image: the fields were filled with very long (225 nm) fibrous molecules, possessing a "head" at one end and

several kinks in the shaft of the fiber (Fig. 1); certainly
the antithesis of Concanavalin A!

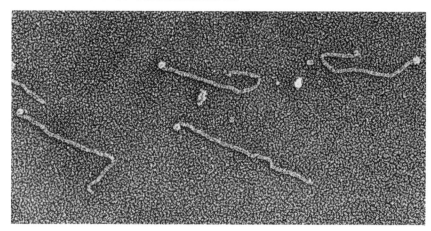

Figure 1. Agglutinin proteins purified from mt⁺
gametes.

Our second important collaborator was Robert Mecham.
His expertise lay in animal extracellular matrix proteins,
and when he subjected material eluted from stacker-gel
slices to amino-acid analysis, he was primed to recognize
the large peak of hydroxyproline that came off the column.
The protein was also rich in serine and, unlike animal
collagens, contained little glycine and no hydroxylysine
(Cooper et al, 1983). These features placed the agglutinin
in the general category of plant hydroxyprotein-rich
glycoproteins (HRGPs).

The third collaborator, James Cooper, was at that time
studying higher- plant HRGPs in the laboratory of Joe
Varner, and he had developed the use of 3,4-dehydroproline,
a non-hydroxylatable proline analog, as an inhibitor of HRGP
synthesis. When we added 3,4-dehydroproline to a mating
mixture of imp-1 (a non-fusing plus mutant) and wild-type
minus gametes (Cooper et al, 1983), agglutinating cells
disengaged with a time course corresponding to turnover of
agglutinin molecules (Snell and Moore, 1980). By contrast,
untreated controls continued to agglutinate for several
hours. Similar results were obtained using the
prolylhydroxylase inhibitor α, α'-dipyridyl, further
documenting the importance of proline hydroxylation for
agglutinin activity.

The realization that Chlamydomonas agglutinins are HRGPs allowed us to place these molecules in an important context, namely, the extensive publications from Keith Roberts' laboratory during the 1970s on the C. reinhardtii cell wall. This series of classic papers, following up initial observations from Derek Lamport's laboratory (Miller et al, 1972), documented that the cell wall of C. reinhardtii is made up of glycoproteins rich in hydroxyproline and serine, with overall compositions strikingly similar to the agglutinins (Catt et al, 1976). In addition, they demonstrated that a subset of wall HRGPs could be selectively extracted with chaotropic agents and made to reassemble in vitro into the same highly ordered crystalline array that they assumed in the native wall (Hills, 1973; Roberts, 1974; Hills et al, 1975; Catt et al, 1978). These findings clearly established that cell wall HRGPs are endowed with molecular recognition properties.

One other chance observation occurred at about this time: an ultracentrifuge tube that had been used for pelleting out cell-wall contaminates of EDTA extracts was inadvertently used for Tetrahymena dynein purification without prior washing. When the "dynein preparation" was viewed by TEM after adsorption to mica flakes, it was found to contain an abundance of fibers, agglutinin-like in general shape but clearly distinct. Unlike agglutinins, these rod-shaped molecules lacked head and kink domains and were arrayed in units that looked like fishbones (Fig. 2). We were thus presented with visual evidence that agglutinin-like wall proteins could indeed assemble into higher-order structures.

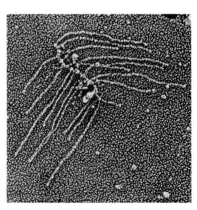

Figure 2. "Fishbone" unit from the W2 layer of the cell wall.

In the summer/fall of 1982, as all these observations and experiments coalesced, we formulated the "central hypothesis" that has guided much of our work for the past 6 years. As originally stated (Cooper et al, 1983): "The self-assembly of Chlamydomonas wall proteins into a crystalline pseudowall (Hills et al, 1975) undoubtedly involves molecular recognition, making it attractive to speculate that the sexual agglutinins have evolved from such structural proteins to perform an intercellular recognition or adhesion function"; and later (Goodenough and Heuser, 1985): "Fibrous hydroxyproline-rich glycoproteins, it is proposed, were initially designed to self-assemble, after secretion, into an extracellular matrix designed to give structural support to the cell. Certain of these glycoproteins then came to be controlled by mating-type loci such that one version of an interacting dyad came to be displayed only on the surface of plus gametic flagella and the other on the surface of minus flagella, so that gametic recognition and adhesion became formally equivalent to wall assembly".

TESTING THE HYPOTHESIS

Most of the work described above focused on the plus agglutinin. In 1983, joined by Patricia Collin-Osdoby, we tackled the minus agglutinin and demonstrated that it is highly homologous to the plus molecule, with subtle differences in morphology and amino acid-composition, the significance of which remains unclear (Collin-Osdoby and Adair, 1985).

Presented with two very similar proteins that display mutual recognition properties, we opted to return to an immunological approach. Using highly purified agglutinin preparations as antigen, we generated a library of monoclonal antibodies in the hope of finding a class which could distinguish between the plus and minus proteins (Adair et al, 1983b). Of the thousands of clonal supernatants screened, from many separate fusions, two clones were found to be agglutinin-specific. Both, however, recognized plus and minus agglutinins equally well. The other 15 well-characterized clones proved to be directed against carbohydrate epitopes shared by all the major cell wall glycoproteins. Of interest was the finding that these epitopes were not present on the 350 kD membrane protein, further establishing an evolutionary link between the agglutinins and cell wall HRGPs (Adair, 1985). In addition,

high-resolution epitope mapping by TEM (Goodenough et al,
1986a) demonstrated that several monoclonal antibodies
recognize homologous domains (e.g., "heads" and "kinks") of
molecules in the agglutinin/wall family. Thus, the
monoclonal antibody approach told us much about the
notion of an HRGP family; however, the ultimate goal of
obtaining a mating-type specific reagent was not achieved.
One of us (WSA) has now launched a project to clone the
agglutinin genes in an effort to identify structural
differences directly.

An important component of the "central hypothesis" is
the idea that agglutinins self-assemble, in the fashion of
cell walls, during flagellar adhesion. It was this line of
thinking that led us into a number of studies on how the
Chlamydomonas cell wall actually assembles (Goodenough and
Heuser, 1985, 1988a, 1988b; Goodenough et al, 1986b; Adair
et al, 1987). Here we could follow a number of leads
provided by the studies of Roberts and co-workers who, as
noted above, had demonstrated the occurrence of crystal
reassembly in vitro. We soon discovered that milligram
quantities of crystal proteins could be extracted from
intact cells, allowing us to purify the three major crystal
HRGPs (GP1, GP2, and GP3) by anion-exchange FPLC and
visualize them by the quick-freeze deep-etch technique
(Goodenough et al, 1986b). Of particular interest was the
finding that GP1, which assembles in a highly ordered array
(Goodenough and Heuser, 1988a), is a 100nm fibrous species
with a knob at one end, and migrates in the stacking gel
during SDSPAGE.

Early work by the Roberts group had also demonstrated
that the salt-insoluble portion of the wall is capable of
"nucleating" assembly of the crystalline wall around itself
(Hills, 1973). We pursued this observation using purified
crystal HRGPs and documented that: 1) crystal HRGPs
polymerize at much lower concentrations when provided with
the salt-insoluble fraction; 2) small foci of
crystallization are "seeded" all over the surface; 3) the
inner sublayer of the crystal (W6A) is formed by the co-
assembly of two HRGPs (GP1 and GP2), while the outer
sublayer (W6B) is a homopolymer of GP1; 4) GP2 and GP3 (but
not GP1) will assemble onto the salt-insoluble matrix; and
5) GP1 will assemble in vitro onto a reconstituted W6A
matrix. In addition, we showed that the salt-insoluble
matrix of Volvox carteri can seed crystallization of C.
reinhardtii proteins and vice-versa, documenting the close
evolutionary relationship between these two species (Adair
et al, 1987).

Taken together, studies of wall-protein polymerization indicate that Chlamydomonas HRGPs have several structural similarities that presumably relate to their assembly properties: heads, tails, and characteristic bends or kinks along the shaft can all serve as recognition/binding sites. In addition, thicker fibers can form by lateral association of individual molecules along their shafts. These properties are familiar to students of animal extracellular matrix, and suggest that plants have evolved analogous mechanisms to achieve similar structural/functional goals.

It is not, unfortunately, yet established which of these principles apply to sexual agglutination, although most have been implicated. Thus, several treatments that inactivate agglutinin function appear to denature or remove the head domain, and the non-head ends serve as binding sites to the flagellar surface (Goodenough et al, 1985), where they presumably associate with intrinsic "agglutinin-binding proteins". In addition, deep-etch images of adhering flagella indicate that the agglutinins associate into a branching meshwork during the adhesion process (Goodenough et al, 1986a and Fig. 3).

Finally, the "central hypothesis" has enabled us to better understand a phenomenon that has long puzzled observers of the Chlamydomonas mating reaction, namely the so-called "requirement for a living cell". Wiese (1969) noted that when he mixed plus and minus gamone preparations they failed to neutralize one another or precipitate out, and several other groups have been unable to agglutinate isolated plus and minus flagella or gluteraldehyde-fixed "corpses", even though each effectively agglutinates living gametes of the opposite mating type. The explanation for this phenomenon appears to lie in a general feature of self-assembly systems, namely the requirement of a critical concentration of reactants for establishment of an assembly nucleus. Thus, if gametic flagella are isolated under conditions where they retain an above-threshold level of agglutinins, flagella-to-flagella adhesion readily occurs in vitro, even if the flagella are pre-fixed in glutaraldehyde (Goodenough, 1986). Therefore, it is not necessary to postulate that adhesion necessitates rearrangements in the plane of the membrane or other "living" activities, only that a sufficient concentration of molecules is available to interact. It is also the case, however, that during the native reaction, complex membrane dynamics are occurring; adhesion complexes move to the flagellar tips (Goodenough and Jurivich, 1978; Goodenough et al, 1980; Musgrave and

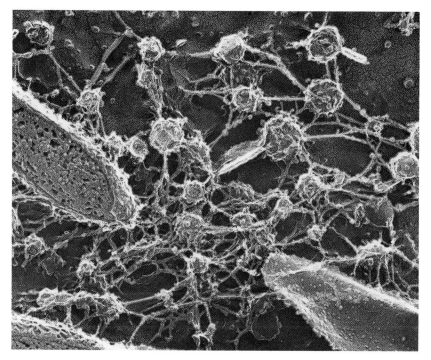

Figure 3. Two agglutinating flagella, interconnected by flagellar membrne vesicles enmeshed in a network of agglutinin adhesion complexes.

van den Ende, 1987), large-scale vesiculation occurs (Fig. 3), and agglutinins are inactivated and replaced (Snell and Moore, 1980; Cooper et al, 1983). Virtually nothing is known about the role that these events may play in the mating reaction.

The key difference between the agglutinins and the cell-wall proteins is, or course, that the agglutinins associate with membranes and their interaction triggers a rapid panoply of mating responses, including cell-wall loss, activation of mating structures, and cell fusion. It has recently been demonstrated that agglutination generates a rise in cAMP levels (Pijst et al, 1984; Pasquale and Goodenough, 1987) and that exogenous presentation of cAMP can induce all the known mating responses in a single mating type (Pasquale and Goodenough, 1987). Therefore, the intrinsic "agglutinin binding proteins" in the flagellar membrane, when subjected to the adhesion stimulus, are somehow able to stimulate adenyl cyclase activity, most likely the cyclase shown to be present in the flagellar membrane (Pasquale and Goodenough, 1987). The next challenge, therefore, is the formulation of a second "central hypothesis" to explain how this adhesion/enzyme-activation coupling system might have evolved.

ACKNOWLEDGEMENTS

The authors warmly acknowledge the friendship and collaboration of the following people who made many contributions to our understanding of Chlamydomonas recognition proteins during the past 10 years: Heidi Apel, Ken Bergman, Eve Caligor, Ronny Cohen, Patricia Collin-Osdoby, Jim Cooper, Pat Detmers, Charlene Forest, Ruth Galloway, Brian Gebhart, Barbara Goldstein, Barry Handler, John Heuser, Carol Hwang, Don Jurivich, David Kirk, Paul Levine, Julie Long, Dianne Mattson, Bob Mecham, Bill Mehard, Val Mermal, Dick Mesland, Brian Monk, Steve Pasquale, Melisse Reichman, Robin Roth, Susan Spath, Scott Steinmetz, Joe Varner.

REFERENCES

Adair WS (1985). Characterization of Chlamydomonas sexual agglutinins. In: Roberts K, Johnson AWB, Lloyd CW, Shaw P, Woolhouse HW, (eds): "The Cell Surface in Plant Growth and Development." J Cell Sci Supp 2:233-260.
Adair WS (1987). Molecular recognition properties of Chlamydomonas HRGPs. In Robinson DG, Starr RC, (eds): "Algal Development: Molecular and Cellular Aspects." Berlin: Springer-Verlag, pp. 90-101.
Adair WS (1988). Organization and in vitro assembly of the Chlamydomonas cell wall. In Varner J, (ed): "Self-assembling Architecture." New York: Alan R. Liss, in press.
Adair WS, Jurivich D, Goodenough UW (1978). Localization of cellular antigens in sodium dodecyl polyacrylamide gels. J Cell Biol 79:281-285.
Adair WS, Monk BC, Cohen R, Goodenough UW (1982). Sexual agglutinins from the Chlamydomonas flagellar membrane. J Biol Chem 257:4593-4602.
Adair WS, Hwang CJ, Goodenough UW (1983a). Identification and characterization of the sexual agglutinin from mating-type plus flagellar membranes of Chlamydomonas. Cell 33:183-193.

Adair WS, Long J, Mehard WB, Heuser JE, Goodenough UW
(1983b). Monoclonal antibodies directed against the
sexual agglutinins of Chlamydomonas reinhardtii. J Cell
Biol 97:93a.
Adair WS, Steinmetz SA, Mattson DM, Goodenough UW, Heuser JE
(1987). Nucleated assembly of Chlamydomonas and Volvox
cell walls. J Cell Biol 105:2373-2382.
Bergman K, Goodenough UW, Goodenough DA, Jawitz J, Martin H
(1975). Gametic differentiation in Chlamydomonas
reinhardtii. II. Flagellar membranes and the
agglutination reaction. J Cell Biol 67:606-622.
Catt JW, Hills GJ, Roberts K (1976). A structural
glycoprotein, containing hydroxyproline, isolated from the
cell wall of Chlamydomonas reinhardii. Planta
131:165-171.
Catt JW, Hills GJ, Roberts K (1978). Cell wall
glycoproteins from Chlamydomonas reinhardtii, and their
self-assembly. Planta 131:165-171.
Collin-Osdoby P, Adair WS (1985). Characterization of the
purified Chlamydomonas minus agglutinin. J Cell Biol
101:1144-1152.
Cooper JB, Adair WS, Mecham RP, Heuser JE, Goodenough UW.
1983. Chlamydomonas agglutinin is a hydroxyproline-rich
glycoprotein. Proc Nat Acad Sci USA. 80:5898-5901.
Goodenough UW (1977). Mating interactions in Chlamydomonas.
In Reissig JL (ed): "Microbial Interactions." London:
Chapman and Hall, pp. 323-350.
Goodenough UW (1985). An essay on the origins and evolution
of eukaryotic sex. In Halvorson H, Monroy A (eds): "The
Origin and Evolution of Sex." New York: Alan R. Liss, pp.
123-140.
Goodenough UW (1986). Experimental analysis of the adhesion
reaction between isolated Chlamydomonas flagella. Exp
Cell Res 166:237-246.
Goodenough UW, Jurivich D (1978). Tipping and mating-
structure activation induced in Chlamydomonas gametes by
flagellar membrane antisera. J Cell Biol 79:680-693.
Goodenough UW, Adair WS (1980). Membrane adhesions between
Chlamydomonas gametes and their role in cell-cell
interactions. In Subtelny S, Wessels NK (eds): "The Cell
Surface: Mediator of Developmental Processes." New York:
Academic Press, pp. 101-112.
Goodenough, UW, Adair WS, Caligor E, Forest CL, Hoffman JL,
Mesland DAM, Spath S (1980). Membrane-membrane and
membrane-ligand interactions in Chlamydomonas matings. In
Gilula NB (ed): "Membrane-Membrane Interactions." New
York: Raven Press, pp. 131-152.

Goodenough UW, Thorner J (1983). Sexual differentiation and mating strategies in the yeast Saccharomyces and the green alga Chlamydomonas. In Yamada KM (ed): "Cell Interactions and Development: Molecular Mechanisms." New York: Wiley and Sons, pp. 29-75.

Goodenough UW, Adair WS, Collin-Osdoby P, Heuser JE. (1985). Structure of the Chlamydomonas agglutinin and related flagellar surface proteins in vitro and in situ. J Cell Biol 101:924-941.

Goodenough UW, Heuser JE (1985). The Chlamydomonas cell wall and its constituent glycoproteins analyzed by the quick-freeze deep-etch technique. J Cell Biol 101:1550-1568.

Goodenough UW, Adair WS, Collin-Osdoby P, Heuser JE (1986a). Chlamydomonas cells in contact. In Gall E, Edelman GM (eds): "Cells In Contact." New York: Wiley and Sons, pp. 111-135.

Goodenough UW, Gebhart B, Mecham RP, Heuser JE (1986b). Crystals of the Chlamydomonas reinhardtii cell wall: polymerization, depolymerization, and purification of glycoprotein monomers. J Cell Biol 103:405-417.

Goodenough UW, Ferris P (1987). Genetic regulation of development in Chlamydomonas. In Loomis W (ed): "Genetic Regulation of Development." New York: Alan R. Liss, pp. 171-189.

Goodenough UW, Heuser JE (1988a). Molecular organization of cell-wall crystals from Chlamydomonas reinhardtii and Volvox carteri. J Cell Sci 90:717-733.

Goodenough UW, Heuser JE (1988b). Molecular organization of the cell wall and cell-wall crystals from Chlamydomonas eugametos. J Cell Sci 90:735-750.

Heuser JE (1983). A method for freeze-drying molecules adsorbed to mica flakes. J Mol Biol 169:155-196.

Hills GJ (1973). Cell wall assembly in vitro from Chlamydomonas reinhardtii. Planta 115:17-23.

Hills GJ, Phillips JM, Gay MR, Roberts K (1975). Self-assembly of a plant cell wall in vitro. J Mol Biol 96:431-441.

Levine RP, Ebersold WT (1960). The genetics and cytology of Chlamydomonas. Annu Rev Microbiol 14:197-216.

Miller DH, Lamport DTA, Miller, M (1972). Hydroxyproline heterooligosaccharides in Chlamydomonas. Science 176:918-920.

Musgrave A, van den Ende H (1987). How Chlamydomonas court their partners. Trends Biochem Sci 12:470-473.

Monk BC, Adair WS, Cohen RA, Goodenough UW (1983). Topography of Chlamydomonas: Fine structure and polypeptide components of the gametic flagellar membrane surface and the cell wall. Planta 158:517-533.

Pasquale SM, Goodenough UW (1987). Cyclic AMP functions as a primary sexual signal in gametes of Chlamydomonas reinhardtii. J Cell Biol 105:2279-2292.

Pijst HLA, van Driel R, Janssens PMW, Musgrave A, van den Ende, H (1984). Cyclic AMP is involved in sexual reproduction of Chlamydomonas eugametos. FEBS Lett. 174:132-136.

Roberts K (1974). Crystalline glycoproteins of algae: Their structure, composition, and assembly. Philos Trans R Soc Lond B Biol Sci 268:129-146.

Snell WJ (1976). Mating in Chlamydomonas: A system for the study of specific cell adhesion. I. Ultrastructural and electrophoretic analysis of flagellar surface components involved in adhesion. J Cell Biol 68:48-69.

Snell WJ (1985). Cell-cell interactions in Chlamydomonas. Ann Rev Plant Physiol 36:287-315.39.

Snell WJ, Moore WS (1980). Aggregation-dependent turnover of flagellar adhesion molecules in Chlamydomonas gametes. J Cell Biol 84:203-210.

van den Ende H (1985). Sexual agglutination in Chlamydomonads. Adv Microb Physiol 26:89-23.

Wiese L (1965). On sexual agglutinatin and mating-type substances (gamones) in isogamous heterothallic Chlamydomonads. I. Evidence of the identity of the gamones with surface components responsible for sexual flagellar contact. J Phycol 1:46-54.

Wiese L (1969). Algae. In Metz CB, Monroy A (eds): "Fertilization: Comparative Morphology, Biochemistry and Immunology." New York: Academic Press, 2:135-188.

Algae as Experimental Systems pages 187–200
© 1989 Alan R. Liss, Inc.

MODULATION OF SEXUAL AGGLUTINABILITY IN CHLAMYDOMONAS
EUGAMETOS (Chlorophyceae)

H. van den Ende, A.M. Tomson, R. Demets, R. Kooijman

Department of Molecular Cell Biology, University of
Amsterdam, The Netherlands

In this report we describe how cells of the biflagellate
unicellular green alga Chlamydomonas eugametos interact with
each other as a prelude to sexual reproduction. We
concentrate on the major recent advances in our understanding
of this process. Reference will be made to the related C.
reinhardtii when appropriate (see Goodenough & Adair, 1989).

Cell-cell interaction is a natural part of the sexual
activity, preparing the cells for cell fusion. In C.
eugametos, as in sperm-egg or pollen-stigma interaction (e.g.
Shapiro, 1987; Chadwick & Garrod, 1986), this preparatory
process involves a physical contact between partner cells and
it is clear that components of the cell surface must play a
decisive role, first to stabilize cellular contact in a
specific way, and second to generate a "signal" by which the
cells are triggered to fuse.

In the sexual process between gametes of Chlamydomonas,
the flagella play a dominating role. They display a mating
type-specific adhesiveness, by which partner cells adhere
together prior to fusion. From the fact that this
adhesiveness is only a transitory property that disappears
when cell fusion is complete, it is obvious that adhesion
between mating cells is not a passive process, but is subject
to regulatory influences from the cell. In this paper it will
be shown that flagellar adhesiveness can be modulated in a
subtle way, by altering the content, the distribution or the
structure of the molecules involved.

THE AGGLUTININS

In the heterothallic C. eugametos, two strains are used with different mating types (mt⁺, mt⁻). In the gamete stage, the cells carry mating type-specific agglutinins at their flagellar surface (Musgrave et al., 1981; Klis et al., 1985; Samson et al., 1987a, b; Adair et al., 1983; Adair, 1985; Goodenough et al., 1985), which are responsible for the mating type-, species- and gamete-specific adhesiveness between mt⁺ and mt⁻ flagella. They are glycoconjugates with a molecular mass of 1200-1300 kD, containing approximately 50% carbohydrate. They are probably extrinsically bound to the flagellar membrane. The question is whether the mt⁺ agglutinin reacts with the mt⁻ agglutinin (unipolar system) or alternatively, whether each of them can function as the ligand for a specific receptor molecule (bipolar system). An argument for the first option is that a monoclonal antibody (mAb) directed against the mt⁻ agglutinin (mAb 66.3), as well as its monovalent Fab fragments, specifically inhibit agglutinin action in vitro, and also block sexual agglutination in vivo (Homan et al., 1988). Thus blocking the action of one of either agglutinins blocks sexual adhesion completely. In the case of a bipolar adhesion system, agglutination might be affected but not completely inhibited, when only one of the agglutinins was inactivated.

TIPPING

Goodenough & Jurivich (1978) observed that when gametes of C. reinhardtii were presented with antiserum raised against the flagellar surface, the antibodies did not bind uniformly over the flagella but were concentrated at the tips. This was considered evidence for the lateral motility of surface receptors in the plane of the membrane that could be directed to the tips of the flagella. Using mAbs against C. eugametos glycoproteins, this observation was confirmed and considerably extended. With the mAb 66.3, mentioned above, it was possible to specifically label the agglutinins on flagella of mt⁻ gametes and to demonstrate that they were redistributed during sexual agglutination (Homan et al., 1987). When gametes were mixed and at various times fixed with glutaraldehyde, it was shown that the distribution of mt⁻ agglutinin originally was evenly distributed over the flagella, but in the course of the agglutination process became concentrated at the flagellar tips. After cell

fusion, when gamete flagella de-agglutinated, this
concentration of the label at the tips also disappeared. The
tip-oriented transport of agglutinins is not a mass flow
phenomenon, since other flagellar glycoproteins, that can be
labeled with various other mAbs, appeared not to migrate
laterally. So in C. eugametos tipping is a specific process,
restricted to the agglutinins and molecules that are
associated with them. One of these is a wheat germ agglutinin
(WGA)-binding protein (Kooijman et al., 1986; unpublished).

Tipping could also be achieved artificially by treating
living mt⁻ gametes with labeled mAb 66.3. The label was then
observed to be first organized into small patches over the
entire length of the flagellar surface and subsequently to
become concentrated at the tips. The conclusion is that the
agglutinins are transported to the flagellar tips as the
result of the cross-linking action of the antibody. This was
confirmed by showing that monovalent antibody fragments were
ineffective in this respect. The mechanism underlying
agglutinin tipping, and more specifically, the transport of
patched agglutinin to the tips, is not known. Since the
process is inhibited by colchicine, it can be envisaged that
the submembranous cytoskeleton in the flagella is involved
(Mesland et al., 1980; Homan et al., 1988). Agglutinin
clustering might establish a microenvironment in the membrane
that facilitates interaction between the agglutinin-
associated intrinsic glycoproteins and the cytoskeleton.

In C. eugametos and C. reinhardtii, this phenomenon
explains how agglutinating flagella become associated by
their tips (e.g. Mesland, 1976), which contributes to the
sorting of cells into pairs (Musgrave et al., 1985), but it
is conceivable that it is also involved in triggering cell
fusion, because it is frequently observed that in a natural
situation only tipped cells will fuse. This is also borne out
by the fact that treating mt⁺ or mt⁻ gametes with mAb 66.3
evokes the same cellular response as does sexual
agglutination with partner cells; monovalent fragments of mAb
66.3 are inactive. Since this reaction can also be obtained
with mAbs that interact with the agglutinins outside the
active site, it seems that it is the cross-linking, resulting
in the formation of patches of agglutinin, which activates
the gametes. Apparently it is irrelevant by what mechanism
patching is elicited. In several other systems receptor
clustering precedes the biological responses, such as the
epidermal growth factor (Schlessinger, 1986), the platelet

fibrinogen receptor (Isenberg et al., 1987) and the complement receptor on human neutrophil cells (Detmers et al., 1987).

MODULATION OF SEXUAL AGGLUTINABILITY

There are several indications that sexual agglutinability in gametes of C. eugametos is modulated. A priori, one may envisage that this is caused by either increasing or decreasing the agglutinin content of the flagella, or by changing the affinity of the agglutinins for their receptors of the opposite mating type. It appears that both mechanisms are operative.

In a light-dark regime, gametes of C. eugametos show a diurnal periodicity in agglutinability (Demets et al., 1987). At the start of the light period, the agglutinability is high, but it declines in the course of the light period, to increase again during the dark. This rhythm persists in continuous dark, while it rapidly damps out in continuous light. This fluctuation can be visualized by immuno-cytochemistry because there is a correlation between agglutinability and the ability to bind mAb 66.3 (Tomson et al., 1988). Bioassays, as well as polyacrylamide gel electrophoresis in SDS, demonstrate a variation in activity and agglutinin abundance, respectively, that parallels the fluctuation in sexual agglutinability. This suggests that the circadian fluctuations in agglutinability of gametes are the consequence of fluctuations in the concentration of agglutinin molecules on the flagellar surface.

A similar mechanism may be inferred with respect to the loss of adhesiveness which is seen after cell fusion. It has been demonstrated by several authors that in agglutinating cells the agglutinin molecules are subject to rapid turnover, due to the inactivation or sloughing of agglutinin molecules and concomitant replenishment of fresh molecules from a cellular pool (Snell & Moore, 1980; Pijst et al., 1984; Saito et al., 1985). We assume that this supply of fresh agglutinins is arrested after cell fusion, resulting in a decline in the content of active agglutinin molecules on the flagella (Musgrave et al., 1985).

SEXUAL AGGLUTINATION IS SELF-ENHANCING

A remarkable property of sexual agglutination in C. eugametos is that it is self-enhancing (Tomson et al., 1986; Demets et al., 1988). This is demonstrated by the following experiments. When gametes of opposite mating type were mixed, an immediate agglutination was observed. By fixing cells at different times during this agglutination process, and testing cells of either mating type by mixing them with live partners in a dilution series, it was found that the adhesiveness of the flagella increased eight- to ten-fold without a preceding lag period (Fig. 1). A maximum value was reached 10-15 min after mixing. Thereafter, the agglutinability of the cells decreased, due to the formation of partially fused cells (vis-à-vis pairs), which are no longer adhesive. Although the rise in adhesiveness took place

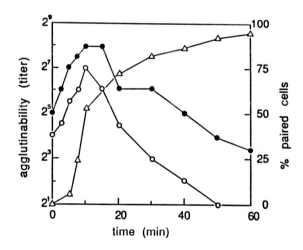

Fig. 1: Agglutinability of C. eugametos during mating: t=0: suspensions of mt+ and mt- gametes mixed. Mating was arrested by adding glutaraldehyde (GA). GA-fixed cell mixtures were washed and tested for reactivity towards living mt+ gametes (giving the agglutinability of the fixed mt- cells, ●—●) and living mt- gametes (idem of the fixed mt+ cells, O—O). Data are mean values of duplicates. From Demets et al. (1988) with permission.

equally in both mating types, the decline did not. The decrease in the mt⁺ cells was more rapid than in the mt⁻ cells. When cells were mixed in 10 mM Tris buffer, pH 9, a condition in which cell fusion but not flagellar adhesion is inhibited, the agglutination reaction persisted for several hours, and then a constant high level of agglutinability was maintained. If the agglutination reaction was disturbed by vigorous shaking, the cells disadhered and the agglutinability in both mating types dropped within minutes to the original level (Fig. 2). This procedure could be repeated several times, which demonstrates that the rise in adhesiveness is a fully reversible, contact-dependent reaction (Demets et al., 1988).

The rise in sexual adhesiveness of flagella of mating cells was paralleled by an increased ability to bind the monoclonal antibody mAb 66.3. When mating cell mixtures were fixed with glutaraldehyde at various times and after washing were presented with mAb 66.3, the antibody, after labeling with gold and application of the silver enhancement procedure, was shown to bind more abundantly to flagella of gametes that had been agglutinating for 15 min than flagella

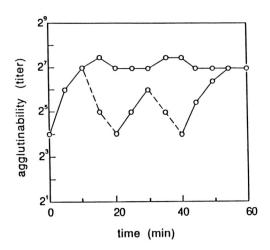

Fig. 2: Increase of agglutinability of gametes mixed at t=0 in 10 mM Tris.HCl, pH 9:-----: cell-cell adhesions severed by shaking. From Demets et al. (1988) with permission.

of gametes that had just been mixed or had fused to form vis-
à-vis pairs. Under the light microscope, places where the
antibody had bound were visible as dark spots at t=0, evenly
distributed over the flagellar surface. Gametes that were
arrested after some minutes of agglutination showed a
significantly stronger labeling. The increase took place
essentially over the entire length of the flagella but most
prominently at the tips. These results show a nice
correlation between the increase in adhesiveness and increase
in antibody-binding during sexual agglutination in mt$^+$ as
well as in mt$^-$ flagella.

The question is how this increase in flagellar
adhesiveness is brought about. It was found that it could
also be evoked by presenting cells with isolated flagella or
isoagglutinin vesicles of the opposite mating type, with a
typical sigmoid dose-response relationship. The response of
the cells depended on the amount of added vesicles. The
increase in agglutinability was never greater than tenfold.
Since the response to the added flagellar membrane vesicles
was mating-type specific and because the main difference
between mt$^+$ and mt$^-$ vesicles is the type of agglutinin they
carry, it was suggested that the gametes were stimulated by
the agglutinins of the opposite mating type. This was
strenghtened by the fact that the rise in agglutinability
could also be realized by adding purified agglutinin of the
opposite mating type. Thus the agglutinins act not only as
ligands/receptors in a binding reaction, but also as
stimulants by which the flagellar adhesiveness of the cells
of the opposite mating type is enhanced. In a mixture of
cells, the mutual stimulation will give rise to a gradual
increase of agglutinability to an maximal level, which as
shown in Fig. 2, can be interrupted when only partially
complete.

These results help to explain a phenomenon, known for a
long time, namely the formation of a specific agglutination
pattern when complementary cell suspensions of high cell
density are mixed ("clumping"; e.g. Brown et al., 1968;
Mesland, 1978). One then observes the rapid formation of a
number of clumps of agglutinating cells; although with time
more and more cells become engaged in agglutination, the
number of clumps does not increase, they just become larger.
Our present interpretation is that cells which are engaged in
sexual contacts, are more adhesive than single, swimming
cells. Therefore, free swimming gametes have a greater chance

of adhering to potential partners in an agglutinating cell
clump than to those that are still swimming free.

Another long-standing problem, namely why it would be
advantageous for gametes to release agglutinin-rich vesicles
("isoagglutinin") into their environment, can be rationalized
(Förster et al., 1956; Wiese, 1984). Contact with
complementary isoagglutinin vesicles makes the cells more
adhesive. At high cell densities, pair formation is not
improved when such vesicles are added, apparently because the
stimulation is already maximal. Only at low densities (below
10^5 cells/ml, when successful cell encounters are infrequent)
pair formation is improved by preincubating the cells with
isoagglutinin of the opposite mating type. Thus the release
of isoagglutinin by gametes of one mating type positively
influences the mating ability of partner cells. So in this
view Wiese and associates (see Förster et al., 1956) were
right when they designated isoagglutinin as a "gamone". It
should be added that isoagglutinin is not always found in
gamete suspensions. Only particular strains, especially mt⁻,
will display a detectable level of production of this
material, which presumably originates by blebbing of the
flagellar membrane. We will argue that this could be due to
an instability of the membrane, caused by local high
concentrations of agglutinin.

What is the molecular background of the sexually induced
increase in adhesiveness of gametes? The most obvious
supposition is that the agglutinin content of the flagella
increases upon physical contact with a partner cell. This was
determined by amputating flagella before and after sexual
stimulation of cells by dibucaine treatment (Thompson et al.,
1974), and after purification extracting them with SDS. The
extracts were assayed for agglutinin content by the charcoal
assay (Musgrave et al., 1981). Mt⁻ flagella, isolated from
gametes after 10 min of agglutination with partner cells (at
7° to slow down the pair formation) showed an increase in
adhesiveness that corresponded with the increase in
adhesiveness of the intact cells. However, these flagella
contained on average only 1.4 times more agglutinin than
flagella from non-stimulated cells. This suggests that the
contact-induced rise in agglutinability and mAb 66.3 binding
is not due to incorporation of additional agglutinin
molecules but rather is the result of increased binding
ability of the agglutinins already present.
When the flagella were detached from the cell bodies by

the more generally applied pH shock procedure, they had almost completely lost their adhesiveness. On extraction, only 30% of the original agglutinin content was recovered from the flagella. This appeared to be due to the release of agglutinin into the medium, both as pelletable material (agglutinin-rich vesicles or aggregates of agglutinin) and as soluble agglutinin. In non-stimulated gametes, in contrast, such a loss was much less severe, about 70% of the agglutinin being retained on the flagella. The implication is that by the contact-induced stimulation of flagellar adhesiveness, the agglutinins become more sensitive to membrane shock, and are readily released into the medium.

There are two circumstantial arguments for the view that the increased adhesiveness is caused by a lateral redistribution of agglutinins in the plane of the membrane. The first argument is the fact that agents interfering with tubulin action, such as colchicine and vinblastine, inhibit contact-induced increase of adhesiveness. In the presence of

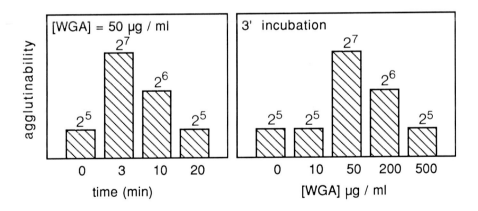

Fig. 3: Increase of agglutinability of mt⁻ cells by the addition of wheat germ agglutinin. Cells were incubated, fixed with GA and assayed (see Fig. 1) after addition of 20 uM N-acetylglucosamine.

these drugs, gametes of opposite mating type adhere temporarily, but stable adhesions do not occur. The adhesiveness of the flagella is not enhanced by contact with partner flagella. Cytochalasin B, a microfilament-inhibiting drug, has no effect. This suggests that the increase of adhesiveness is dependent on the action of the microtubular cytoskeleton in the flagella. As discussed above, tipping, which is seen as the ultimate consequence of agglutinin clustering in the flagellar membrane, is also susceptible to the action of tubulin-blocking drugs. A second argument is that a treatment of gametes with wheat germ agglutinin (WGA) results in enhanced adhesiveness of the flagella that is completely comparable with contact-induced increase of adhesiveness (Fig. 3). Since WGA binds to a membrane component which is associated with agglutinin (it is co-distributed with agglutinin during tipping), and because it is multivalent, the conclusion is that it causes a clustering of agglutinins similar to what happens on sexual contact, with a concomitant increase in the affinity between the flagella, but by a different mechanism.

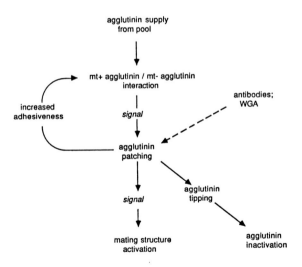

Fig. 4: Proposed scenario for activating gametes of C. eugametos for fusion.

In summary, we propose that preparing gametes for sexual fusion proceeds along the following scenario (Fig. 4): heterotypic agglutinin-agglutinin binding results in a signal which induces a cytoskeleton-driven clustering of agglutinin molecules; this results in a) an increased affinity between the two flagella, for example by a cooperative increase of the agglutinin-agglutinin binding constant or by an aggregation-induced allosteric change of the molecule; b) the generation of a second (or stronger) signal triggering cellular responses like mating structure activation; and c) a cytoskeleton-driven transport of agglutinin clusters to the flagellar tips.

We speculate that the local concentrations of agglutinin molecules, together with associated proteins, results in a local instability of the flagellar membrane, resulting in the ready detachment of agglutinin-enriched vesicles. This would explain the result described by Snell et al. (1986), who used Sepharose beads coated with a mAb against the mt$^+$ agglutinin of C. reinhardtii. When such beads were incubated with mt$^+$ gametes, these adhered specifically to the beads, but also appeared to transfer agglutinin from the flagellar membrane to the bead surface, which consequently became adhesive for mt$^-$ gametes. A similar result was obtained with C. eugametos by presenting mt$^-$ gametes with charcoal particles, coated with mt$^+$ agglutinin. Initially, the mt$^+$ gametes adhered to the particles, but soon the particles lost their adhesiveness, while at the same time they became adhesive towards mt$^+$ cells. We presume that the mt$^-$ gametes were stimulated by the mt$^+$ agglutinin on the charcoal and reacted by an increased clustering of their agglutinin on the flagella, but thereby lost much of their agglutinin to the adsorbent. The charcoal became consequently loaded with both types of agglutinins. The mt$^+$ gametes did not exhibit this phenomenon, presumably because they are more resistant to loss of agglutinin in this type of experiment.

Pursuing this line of reasoning, we can imagine that agglutinin clustering during sexual interaction of mt$^+$ and mt$^-$ gametes eventually results in a concentration of agglutinin at the flagellar tips, from where it tends to be shed in the form of membrane vesicles. This could explain the increased agglutinin turnover during sexual agglutination, as was described above. Of course, this view must be consolidated by directly demonstrating the release of mt$^+$ and mt$^-$ agglutinins into the medium containing the interacting gametes.

REFERENCES

Adair WS (1985). Characterization of Chlamydomonas sexual
  agglutinins. J Cell Sci Suppl 2: 233-260.
Adair WS, Hwang C, Goodenough UW (1983). Identification and
  visualization of the sexual agglutinin from the mating-type
  plus flagellar membrane of Chlamydomonas. Cell 33:183-193.
Brown RM, Johnson C, Bold HC (1968). Electron and phase-
  contrast microscopy of sexual reproduction in Chlamydomonas
  moewusii. J Phycol 4: 100-120.
Chadwick CM, Garrod DR (eds) (1986). "Hormones, Receptors and
  Cellular Interactions in Plants. Intercellular and
  Intracellular Communication. Cambridge: University Press,
  Vol. 1, xii + 375 pp.
Demets R, Tomson AM, Stegwee D, van den Ende H (1987).
  Control of the mating competence rhythm in Chlamydomonas
  eugametos. J Gen Microbiol 133: 1081-1088.
Demets R, Tomson AM, Homan, WL, Stegwee, D, van den Ende H
  (1988). Cell-cell adhesion in conjugating Chlamydomonas
  gametes: a self-enhancing process. Protoplasma 145: 27-36.
Detmers PA, Wright SD, Olsen E, Kimball B, Kohn ZA (1987).
  Aggregation of complement receptors on human neutrophils in
  the absence of ligand. J Cell Biol 105: 1137-1145.
Förster H, Wiese L, Braunitzer G (1956). Über das
  agglutinierend wirkende Gynogamon von Chlamydomonas
  eugametos. Z Naturf 11b: 315-317.
Goodenough UW, Jurivich D (1978). Tipping and mating
  structure activation in Chlamydomonas gametes by flagellar
  membrane antisera. J Cell Biol 79: 680-693.
Goodenough UW, Adair WS, Collin-Osdoby P, Heuser JE (1985).
  Structure of the Chlamydomonas agglutinin and related
  flagellar surface proteins in vitro and in situ. J Cell
  Biol 101: 924-941.
Goodenough UW, Adair WS (1989). Recognition proteins of
  Chlamydomonas reinhardtii (Chlorophyceae). In Coleman AW,
  Goff LJ, Stein-Taylor JR (eds): "Algae as Experimental
  Systems", New York: Alan R Liss, Inc pp171-185
Homan WL, Musgrave A, de Nobel H, Wagter R, de Wit D, Kolk A,
  van den Ende H (1988). Monoclonal antibodies directed
  against the binding site of Chlamydomonas eugametos
  gametes. J Cell Biol 107: 177-189
Homan WL, Sigon C, van den Briel W, Wagter R, de Nobel H,
  Mesland DAM, van den Ende H (1987).Transport of membrane
  receptors and the mechanics of sexual cell fusion in
  Chlamydomonas eugametos. FEBS Lett. 215: 323-326.
Isenberg WM, McEver RP, Phillips DR, Shuman MA (1987). The

platelet fibrinogen receptor: an immunogold-surface replica
study of agonist-induced ligand binding and receptor
clustering. J Cell Biol 104: 1655-1663.
Klis FM, Samson MR, Touw E, Musgrave A, van den Ende H
(1985). Sexual agglutination in the unicellular green alga
Chlamydomonas eugametos. Identification and properties of
the mating type plus agglutination factor. Plant Physiol
79: 740-745.
Kooijman R, Elzenga TJM, de Wildt P, Musgrave A, Schuring F,
van den Ende H (1986). Light dependence of sexual
agglutinability in Chlamydomonas eugametos. Planta 169:
370-378.
Kooijman R, de Wildt P, Homan WL, Musgrave A, van den Ende H
(1988). Light affects flagellar agglutinability in
Chlamydomonas eugametos by modification of the agglutinin
molecules. Plant Physiol 86: 216-223.
Mesland DAM (1976). Mating in Chlamydomonas eugametos. A
scanning electron microscopical study. Arch Microbiol
109:31-35.
Mesland, DAM (1978). Sexual cell interaction in Chlamydomonas
eugametos. Doctoral Dissertation. University of Amsterdam.
Mesland DAM, Hoffman JL, Caligor E, Goodenough UW (1980).
Flagellar tip activation stimulated by membrane adhesions
in Chlamydomonas gametes. J Cell Biol 84:599-617.
Musgrave A, van Eijk E, te Welscher R, Broekman R, Lens P,
Homan WL, van den Ende H (1981). Sexual agglutination
factor from Chlamydomonas eugametos. Planta 153: 362-369.
Musgrave A, de Wildt P, Schuring F, Crabbendam K, van den
Ende H (1985). Sexual agglutination in Chlamydomonas
eugametos before and after cell fusion. Planta 166: 234-
243.
Pijst HLA, Ossendorp FA, van Egmond, P, Kamps A, Musgrave A,
van den Ende H (1984). Sex-specific binding and
inactivation of agglutination factor in Chlamydomonas
eugametos. Planta 160: 529-535.
Saito T, Tsubo Y, Matsuda Y (1985). Synthesis and turnover of
cell body-agglutinin as a pool for flagellar surface-
agglutinin in Chlamydomonas reinhardtii. Arch Microbiol
142: 207-210.
Samson MR, Klis FM, Crabbendam KJ, van Egmond P, van den Ende
H (1987a). Purification, visualization and characterization
of the sexual agglutinins of the green alga Chlamydomonas
moewusii yapensis. J Gen Microbiol 133: 3183-3191.
Samson MR, Klis FM, Homan WL, van Egmond P, Musgrave A, van
den Ende H (1987b). Composition and properties of the
sexual agglutinins of the flagellated green alga

Chlamydomonas eugametos. Planta 170: 314-321.
Schlessinger J (1986). Allosteric regulation of the epidermal growth factor receptor kinase. J Cell Biol 103: 2067-2072.
Shapiro BM (1987). The existential decision of a sperm. Cell 49: 293-294.
Snell WJ, Moore WS (1980). Aggregation-dependent turnover of flagellar adhesion molecules in Chlamydomonas gametes. J Cell Biol 84: 203-210.
Snell WJ, Kosfitzer MG, Clausell A, Perillo N, Imam S, Hunnicutt, G (1986). A monoclonal antibody that blocks adhesion of Chlamydomonas mt$^+$ gametes. J Cell Biol 103: 2449-2456.
Thompson GA, Baugh LC, Walker LF (1974). Non lethal deciliation of Tetrahymena by a local anesthetic and its utility as a tool for studying cilia regeneration. J Cell Biol 61: 253-257.
Tomson AM, Demets R, Homan WL, Stegwee D, van den Ende H (1988). Endogenous oscillator controls mating receptors in Chlamydomonas. Sexual Plant Repr 1, 46-50.
Wiese L (1984). Mating systems in unicellular algae. Encycl. Plant Physiol, NS 17: 238-284.

Algae as Experimental Systems pages 201–213
© **1989 Alan R. Liss, Inc.**

THE ROLE OF PHEROMONES IN SEXUAL REPRODUCTION OF BROWN
ALGAE

Dieter G. Müller

Fakultät für Biologie der Universität,
D-7750 Konstanz, Federal Republic Germany

Sexual reproduction in eucaryotes comprises plasmogamy
and meiosis. It requires specific cooperation of two complementary partners to form the zygote. Many marine brown algae
demonstrate very distinct attraction of male to female gametes, which implies the presence of attractants secreted into
the surrounding sea water. The demonstration of Fucus eggs
with halos of attracted sperm is a long-standing classroom
experiment in marine biology. Over the years many workers
tried to identify such attractants without success. The
arrival of advanced methods in analytical chemistry marked
a breakthrough, and cooperation between biologists and
anylytical chemists made possible the analysis of biological
materials in nanogram quantities by gas chromatography and
mass spectrometry. Starting from 1967 the fertilization
pheromones in marine brown algae have been extensively investigated (Maier & Müller, 1986). The substances discussed
here convey information between cells through water as an
external medium and thus justify application of the term
pheromone, which was introduced by Karlson and Lüscher
(1959). In this review I will try to give a short survey of
various aspects of the present state of knowledge on brown
algal pheromones.

In this group of algae sexual fusion takes place between iso- or anisogametes, or between eggs and spermatozoids.
After settling down on the substrate, a female gamete becomes attractive to the respective male gametes until the
zygote is formed. In many instances the secretions from
female gametes escape into the air and can be sensed as a

pleasant fragrance with the human nose. Egg secretions in the orders Laminariales, Desmarestiales and Sporochnales have two effects: liberation of spermatozoids from the antheridium and their subsequent attraction to the egg.

ISOLATION AND STRUCTURE DETERMINATION

The odoriferous character of the brown algal pheromones indicates they are lipophilic and highly volatile. This property is used for isolation by stripping them from the aqueous phase with air. The apparatus is a closed-circuit glass system with intense water-air-exchange maintained by a miniature air pump. Volatile organic material is adsorbed from the air onto a filter of activated carbon. Simulation experiments with recovery of synthetic pheromones showed that considerable losses occur. The best recovery rates of ectocarpene was 17%, and a strong dependence on the volume of the extraction vessel was found: recovery rates decreased with increasing volume (Müller & Schmid, 1988).

After elution with dichloromethane, a typical extract contains microgram quantities of egg secretions. It can be directly subjected to GC-MS analysis and suitable micro-reactions such as hydrogenation. The information gained with these techniques is not normally sufficient for structure determination. This difficulty is overcome by synthesis of possible "candidate" structures until a complete match of Kovats indices on different capillary GC-columns and perfect identity of mass spectra is achieved. As the last step, bioassays confirm the biological activity of the synthetic structure. In a few instances it has recently been possible to use enantiomer-specific HPLC techniques to determine the chirality of pheromones (Boland et al., 1985, 1987a).

THE KNOWN BROWN ALGAL PHEROMONES

Within the past two decades ten different pheromone molecules have been isolated and identified from more than fifty species of brown algae. All these molecules are olefinic hydrocarbons. With the exception of fucoserratene, which is an octatriene (C_8), all substances are C_{11} skeletons. They include open chains or cyclopropane- cyclopentene- or cycloheptadiene derivatives. So far only one structure with an epoxidic oxygen hetero-atom has been encountered (Fig. 1; Maier & Müller, 1986).

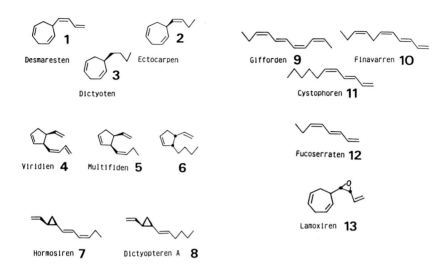

Figure 1. Structural formulae showing sexual pheromones of brown algae identified to date: compounds 6, 8, 9 are secretions with no detectable biological activity.

PHEROMONE BOUQUETS

High-resolution gas chromatography reveals that most if not all, secretions from female brown algal gametes contain a mixture of compounds. In addition to the main fraction, which normally represents the active pheromone, traces of chemically related by-products are encountered. This finding is consistent with the notion of fatty acid precursors processed by enzymes with limited specificity. At present there is no conclusive answer to the question, whether this bouquet-character is merely accidental or whether it is biologically relevant.

BIOSYNTHESIS AND BIOASSAYS

Algal gametes are specialized cells and available in only limited quantities. This has so far excluded studies on biosynthesis of pheromones. However, the essential oil of a terrestrial plant, the South African composite <u>Senecio</u>

isatideus, is dominated by ectocarpene. Precursor studies
with this plant give conclusive evidence that ectocarpene
is derived from dodecatrienoic acid. From this and slightly
modified precursor molecules, the entire family of brown
algal pheromones can easily be derived (Boland & Mertes,
1985). At present it remains uncertain, whether this pathway also applies to algal gametes.

Efforts toward isolation and structure determination
require that the biological activity of various compounds
is evaluated. The administration of low-molecular, volatile lipophilic compounds in an aqueous medium requires
special procedures. One type of a qualitative experiment is
the adsorption of the compound to be assayed onto porous
particles of silica. When brought in contact with male
gametes, they act as attractant sources. Quantitative data
can be obtained with a high-density fluorocarbon solvent
which is biologically inert and insoluble in water. Micro-
droplets of this solvent with known concentrations of an
attractant are placed into suspensions of active male
gametes. After a few minutes exposure the higher density of
cells over experimental droplets, when compared to a blank
consisting of pure solvent, illustrates the attractive
potential. Dark-field photomicrography documents the gamete
distribution and makes possible numerical evaluation and
statistical treatment of the data (Fig. 2). In this type of
experiment it is necessary to know the partition coefficient
of the attractant, which expresses the ratio of concentrations in the solvent and in the surrounding water. These
coefficients range from several thousand for highly lipophilic molecules like ectocarpene to about 30 in more
hydrophilic molecules like lamoxirene.

THRESHOLD CONCENTRATIONS

Bio-assays permit determination of the lowest pheromone concentration that induces a statistically significant
response in male gametes. Maximum sensitivity has been
found in Cutleria multifida with a threshold concentration
of 2×10^{-11} M. Calculations incorporating diffusion con-
stants of the pheromones and geometrical parameters of the
gametes show, that pheromone molecule encounter rate values
as low as 100 molecules per 80 ms are sufficient for chemotactic responses of a male gamete. This is a remarkably low
value, similar to those reported for pheromone perception
in insects (Boland, 1987).

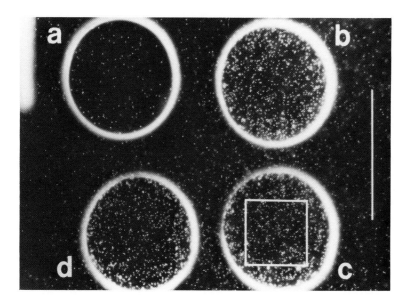

Figure 2. Quantitative bioassay with four micro-droplets of the fluorinated solvent FC-78: a) is blank; b, c, d contain multifidene which attracts spermatozoids of Cutleria multifida. Scale is 1 mm.

SECRETION RATES

A detailed study of pheromone secretion in Ectocarpus siliculosus gave the following results: after a brief motility period of about 30 min female gametes settled down on the substrate. Ectocarpene could be extracted for about 7 h. During this time each cell was calculated to secrete 10^5 molecules x s^{-1}. The total pheromone secretion by a female gamete corresponded to 0.3% of its fresh weight. Similar values between 0.2 and 0.4% were found for the larger eggs of several kelps (order Laminariales; Müller & Schmid, 1988).

RECEPTOR SPECIFICITY

Two model systems, Cutleria and Laminaria were studied to characterize the degree of specificity of pheromone recognition. Series of specifically modified synthetic

pheromone analogues were assayed for their biological
effects. In summary, any alteration of the pheromone molecule resulted in a reduction of its biological effectiveness. The most important features were found to be its
overall shape and the correct spatial arrangement of double
bonds (Boland et al., 1981; Maier et al., 1988). This suggests that the pheromone is recognized at the receptor site
by dipole-dipole interactions and close spatial fitting.

TAXONOMIC SPECIFICITY

Ectocarpene has been found in Ectocarpus siliculosus,
E. fasciculatus (Ectocarpales), Sphacelaria rigidula
(Sphacelariales) and Adenocystis utricularis (Dictyosipho-
nales). Thus, it is characteristic for the genus Ectocarpus,
but also found in species from two other orders (Müller et
al. 1985a). Cladostephus spongiosus,(Sphacelariales) produces desmarestene, which is also known from Desmarestia
(Desmarestiales; Müller et al., 1986). Lamoxirene has been
found in all members of the families Laminariaceae, Alaria-
ceae and Lessoniaceae in the order Laminariales (Müller &
Maier, 1985). Three species of the genus Fucus secrete
fucoserratene, whereas other members of this order, as
Ascophyllum and Cystophora have their own genus-specific
attractants (Maier & Müller, 1986). Hormosirene has a very
broad distribution, being found in two families of Fucales
and three species of Durvillaea (Durvillaeales), as well as
in two genera of Scytosiphonales (Scytosiphon, Colpomenia;
Müller et al., 1985b). This rather broad pattern is con-
trasted by the situation in the genus Chorda (family Chor-
daceae, order Laminariales). Here the pheromone of Chorda
tomentosa is multifidene, whereas C. filum has a different
pheromone, which is not fully characterized to date (Maier
et al., 1984). This finding is in accordance with taxonomic
considerations, which suggest that the two species of the
genus Chorda should be separated (Maier, 1984). There is at
the moment no clear scheme which can reconcile the distribution pattern of pheromone molecules with taxonomy.

PHEROMONES AND RELATED COMPOUNDS IN VEGETATIVE THALLI

The genus Dictyopteris (Dictyotales) is well known for
its sweet odor when taken out of the sea, and its secretion
products have been studied in detail (Moore, 1977; Boland &
Müller, 1987). Among more than a dozen different compounds
there are at least four which act as pheromones in other

species. These substances are secreted by vegetative thalli as well as sporophytes and gametophytes. Clearly, there is no connection to sexual reproduction, and the significance of this phenomenon is not understood. A similar instance occurs in Giffordia mitchellae (Ectocarpales), which secretes large amounts of giffordene, a conjugated undeca-tetraene from sterile plants as well as from sporophytes and gametophytes, without an obvious function (Boland et al., 1987b).

UNRESOLVED CASES

The situation in Giffordia mitchellae is puzzling, because microscopical observation shows a dramatic attrac-tion of spermatozoids by settled female gametes (Fig. 3). Giffordene has no attractive potential, and extraction of volatile materials from gametes has not revealed the sperm attractant so far. In Himanthalia elongata (Fucales) on North European coasts, large batches of eggs have been repeatedly subjected to the extraction procedures for vola-tile compounds with negative results. In this species, microscopic evidence does not support the attraction of sperm by the eggs. Likewise, several efforts to isolate a sperm attractant in Sargassum muticum (Fucales) have failed. This species is monoecious, and fertilization takes place within a gelatinous ring that is formed in the fertile area of the receptacle. The spermatozoids do not normally leave this area, and fertilization in this species may be accomplished without sperm chemotaxis.

PHEROMONE PERCEPTION AND SPERMATOZOID RELEASE

Three orders within the class Phaeophyceae are charac-terized by microscopic gametophytes, oogamy and unicellular antheridia: Laminariales, Desmarestiales, Sporochnales. In all three orders, sperm release from antheridia is mediated by a pheromone secreted from freshly released eggs. Eight to ten seconds after perception of the pheromone, the spermatozoids are expelled explosively. Studies in Laminaria digitata showed that the interior of the antheridium is under pressure due to the hydration of mucopolysaccharides. The living spermatozoid is actively involved in its own release. The pheromone has to cross the cell wall and enter the spermatozoid's plasma membrane. Interaction with the pheromone receptor initiates a cascade of events that after

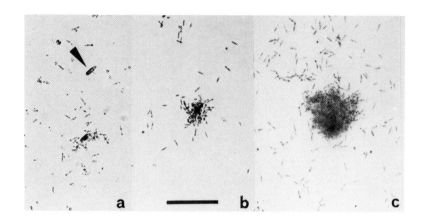

Figure 3. Spermatozoids and female gametes of <u>Giffordia</u>
<u>mitchellae</u>: a) free-swimming female gametes are unattrac-
tive (arrowhead); b, c) settled cells attract large numbers of spermatozoids. b) is 4 min; c) is 8 min after
settlement. Scale is 100 μm.

a few seconds results in the enzymatic softening of a predetermined annular zone at the apex of the antheridium
(Fig. 4). The internal pressure then ejects the spermatozoid, which, after another 2 sec begins its locomotory
activity. The release process is dependent on extracellular
cations, especially Ca^{2+} and Na^+, and experimental results
with Ca-antagonists indicate that transient membrane
fluxes and Ca^{2+} as a second messenger may be involved
(Maier, 1982).

PHEROMONE PERCEPTION AND CHEMOTAXIS

The pheromones effecting the liberation of spermatozoids continue to attract them to the egg cells (Fig. 5).
Tracking of <u>Laminaria</u> spermatozoids with video techniques
indicates that the approach to the egg is more or less
straight - forward up-gradient, and thus complies with the
defination of positive chemotaxis (Fig. 6b). Other instances of chemo-accumulation by attractants show different
cellular responses. In <u>Fucus</u>, spermatozoids seem to enter

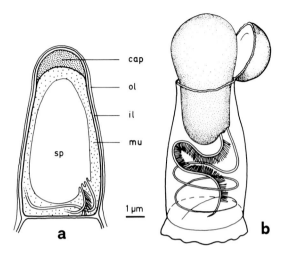

Figure 4. Laminaria digitata. a) longitudinal section
through antheridium with cap. 01 - outer wall; il - inner
wall; mu - mucopolysaccharides surrounding the spermatozoid
(sp). b) spermatozoid during release process.

the pheromone halo around an egg cell randomly and are then
trapped in the vicinity of the egg. When they move away
from the egg and hit a critical threshold concentration of
the pheromone, they execute an abrupt 180° turn (Fig. 6c;
Maier & Müller, 1986). This cellular behaviour is termed a
step-down chemophobic reaction. A third type of pheromone-
mediated sexual approach has been studied in detail in
Ectocarpus siliculosus (Geller & Müller, 1981). Male gametes show a pronounced contact response, which causes them
to move along the substrate with reduced speed. In this
situation the asymmetry of the motile gamete's flagellation
results in elongated clockwise, loop-like tracks when
viewed in the microscope. Upon perception of the pheromone
the contact response is elicited and the flagellar activity
pattern is dramatically altered. Normally, the rear flagellum is passive. Under pheromone stimulation it responds
with violent beats. Similar to a ship's rudder these impul-
ses cause a 180° hairpin-turn or a tight circle.

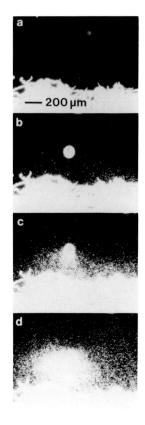

Figure 5. <u>Alaria</u> <u>esculenta</u>. Lower edge in a) is mature male gametophyte. b) introduction of a silica particle impregnated with lamoxirene. c, d) release and attraction of spermatozoids.

The posterior flagellum's activity is directly proportional
to the pheromone concentration, and the result is trapping
of the male gamete in the immediate vicinity of the female
pheromone source by circling and repeated turning reactions.
Eventually, the male gamete establishes contact with the
surface of the female cell, and plasmogamy follows (Fig.
6a). This type of locomotory pattern is termed chemo-
thigmo-klinokinesis, which is best demonstrated by cine-
micrographic presentation (Müller, 1982).

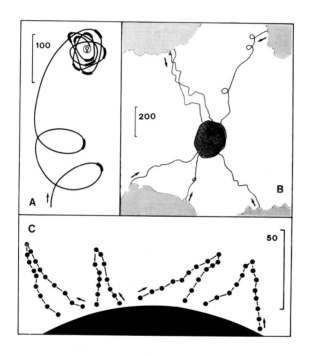

Figure 6. Different types of gamete approach in brown algae
A. Chemo-thigmo-klinokinesis in Ectocarpus siliculosus:
emphasized parts of male tracks indicate periods of hind
flagellum beat. B. Laminaria digitata: impregnated silica
particle as pheromone source in center with tracks of
individual spermatozoids. C. Fucus spiralis: return-
responses of individual spermatozoids near a fluorocarbon-
droplet containing fucoserratene. Scale in µm.

FINAL REMARK

Present studies include the analysis of additional
brown algal pheromone systems. In selected cases such as
Fucus and Laminaria pheromone-receptor interactions are
studied and first attempts to characterize the receptor
sites in spermatozoids are in progress. Laboratory cultures
of Dictyopteris will be used to compare the biosynthetical
pathways for C_{11} olefines in terrestrial plants and brown
algae. The Phaeophyceae offer a unique chance to study
cellular communication as a fundamental biological phenomenon with many variations in a relatively homogeneous
taxonomic group.

REFERENCES

Boland W (1987). Chemische Kommunikation bei der sexuellen
Fortpflanzung mariner Braunalgen. Biol Zeit 17:176-185.
Boland W, Flegel U, Jordt G, Müller DG (1987a). Absolute
configuration and enantiomer composition of hormosirene.
Naturwissenschaften 74:448-449.
Boland W, Jacoby K, Jaenicke L, Müller DG, Fölster E (1981).
Receptor specificity and threshold concentration in
chemotaxis of the Phaeophyte Cutleria multifida. Z
Naturforsch 36c:262-271.
Boland W, Jaenicke L, Müller DG, Gassmann G (1987b).
Giffordene, 2Z, 4Z, 6E,8Z-undecatetraene, is the
odoriferous principle of the marine brown alga Giffordia
mitchellae. Experientia 43:466-467.
Boland W, Mertes K (1985). Biosynthesis of algal pheromones.
A model study with the composite Senecio isatideus. Eur
J Biochem 147:83-91.
Boland W, Müller DG (1987). On the odor of the Mediterra-
nean seaweed Dictyopteris membranacea: New C_{11} hydro-
carbons from marine brown algae III. Tetrahedron Letters
28:307-310.
Geller A, Müller DG (1981). Analysis of the flagellar beat
pattern of male Ectocarpus siliculosus gametes
(Phaeophyta) in relation to chemotactic stimulation by
female cells. J Exp Biol 92:53-66.
Karlson P, Lüscher M (1959). "Pheromones": a new term for a
class of biologically active substances. Nature 183:
55-56.
Maier I (1982). New aspects of pheromone-triggered spermatozoid release in Laminaria digitata (Phaeophyceae).
Protoplasma 113:137-143.

Maier I (1984). Culture studies of Chorda tomentosa (Phaeophyta, Laminariales). Br Phycol J 19:95-106.

Maier I, Müller DG (1986). Sexual pheromones in algae. Biol Bull 170:145-175.

Maier I, Müller DG, Gassmann G, Boland W, Marner F-J, Jaenicke L (1984). Pheromone-triggered gamete release in Chorda tomentosa. Naturwissenschaften 71:48.

Maier I, Müller DG, Schmid C (1988). Pheromone receptor specificity and threshold concentrations for spermatozoid release in Laminaria digitata. Naturwissenschaften 75: 260-263.

Moore RE (1977) Volatile compounds from marine algae. Acc Chem Res 10:40-47.

Müller DG and Inst Wiss Film (1982). Pheromone effects in fertilization of brown algae. Film C 1424 des Instituts für den Wissenschaftlichen Film Göttingen.

Müller DG, Boland W, Jaenicke L, Gassmann G (1985a). Diversification of chemoreceptors in Ectocarpus, Sphacelaria and Adenocystis (Phaeophyceae) Z Naturforsch 40c:457-459.

Müller DG, Clayton MN, Gassmann G, Boland W, Marner, F-J, Schotten T, Jaenicke L (1985b). Cystophorene and hormosirene, sperm attractants in Australian brown algae. Naturwissenschaften 72:97-99.

Müller DG, Clayton MN, Meinderts M, Boland W, Jaenicke L (1986). Sexual pheromone in Cladostephus (Sphacelariales, Phaeophyceae). Naturwissenschaften 73:99.

Müller DG, Maier I (1985). Survey on sexual pheromone specificity in Laminariales (Phaeophyceae). Phycologia 24:475-477.

Müller DG, Schmid C (1988). Qualitative and quantitative determination of pheromone secretion in female gametes of Ectocarpus siliculosus (Phaeophyceae). Biol Chem Hoppe-Seyler 369:647-653.

Algae as Experimental Systems pages 215–230
© 1989 Alan R. Liss, Inc.

THE *CHLAMYDOMONAS* (CHLOROPHYCEAE) EYE AS A MODEL OF CELLULAR
STRUCTURE, INTRACELLULAR SIGNALING AND RHODOPSIN ACTIVATION

Kenneth W. Foster and Jureepan Saranak

Department of Physics, Syracuse University,
Syracuse, New York 13244-1130

"*Chlamidomonas. Animal e familia Volvocinorum, sine
cauda, sed ocello ...*" (Ehrenberg, 1838). Single celled
organisms such as *Chamydomonas* and humans occupy the same
small corner of the evolutionary tree (Smyth *et al.*, 1988;
Sogin *et al.*, 1986); however, *Chlamydomonas* has the microbial
advantages (small genome, haploid genetics, plate colonies,
rapid and easy growth, short generation time) for many studies at the molecular and cellular level. Three questions are
addressed here: 1) How is the precise architecture of a cell
coded and constructed? 2) How are multiple signals handled
within the cell? 3) How is a sensory receptor protein activated and its selectivity regulated? We have investigated
these questions by studying the eye and rhodopsin photopigment of *Chlamydomonas*.

CELLULAR ARCHITECTURE AND THE EYE OF *CHLAMYDOMONAS*

The structure of eyes is a intriguing design problem in
development. By analogy with man-made devices for receiving
and transmitting microwaves, eyes are *antennas* for reception
of light waves. They are of the same order in size as the
light waves with which they interact and thus their optics is
dominated by interference and diffraction. In humans an eye
consists of an antenna array of rod and cone cells at the
focus of a lens with the retina and brain doing the signal
processing. In a cone cell rays of light travel within a
narrow angle of the cell's optical axis in order to be detected. The cone's directivity (sensitivity to light as a
function of the direction the light is coming from) is due to
its dielectric antenna properties (di Francia, 1949). Phototactic algae solve the problem of finding the light direction
by scanning their environment with a "single" eye that is
sensitive to the light in the environment. In *Chlamydomonas*

its eye optimizes light coupling by constructive interference of the incident light to an overlying photopigment (Foster & Smyth, 1980). To meet this physical requirement precise shape and dimensionality, material composition and location within the cell are necessary. Currently most researchers feel that there is no specific code for structure, but rather a structure is the unfolding of a developmental pattern built into the timing of gene expression and the interactions of components (Harold, 1986; Poole and Trinci, 1987), but is this also true for eyes.

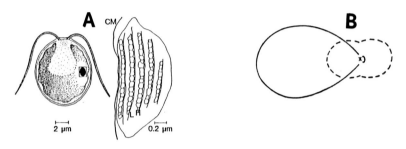

Figure 1. Schematic diagram of *Chlamydomonas* cell and eye (A) showing quarter-wave stack of high refractive index (H) carotenoid globules alternating with low refractive index (L) aqueous layers (CM, cell membrane) and eye directivity (B) with (solid) and without (dashed) eye structure.

In the *Chlamydomonas* eye, the photopigment is located in a membrane overlying a neighboring high refractive index layer at an optical path distance of one-quarter the wavelength of light. A pile of these alternating high and low refractive index layers [1.9 and 1.35 respectively at 2.30 eV (540 nm)] each with an optical path of one-quarter wavelength make up a quarter-wave **stack** (Fig. 1A) as in a laser mirror. A high refractive index is achieved by taking advantage of the fact that the maximum refractive index occurs on the low energy side of strongly absorbing carotene pigments (maxima at 2.75 and 2.53 eV, 450 and 490 nm respectively) and by packing the carotene to a high density since refractive index is also proportional to concentration of pigment. Light entering through the cell is reflected back before reaching the pigment giving the eye monopolar directivity. The layered design also narrows the field of view in the forward (outward) direction when it is non-absorptive (Fig. 1B). This specific design is optimized for a rhodopsin as shown by the match between the intensity focused on the receptor (Fig. 2A)

and the absorption of its rhodopsin which has a maximum at
2.46 eV (503 nm) (Foster *et al.*, 1984).

Figure 2. Spectral coupling to
the receptor (A) and modulation
contrast (B). Solid line,
actual multilayer structure;
dashed line, assumes layers
squeezed together.

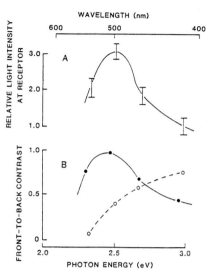

    The cell rotates left-
handedly about 2 Hz as it
swims with its two flagella
forward and beating in a
ciliary beat (Foster & Smyth,
1980). The eye which is placed
on the side of the cell looking
outward and laterally is then
part of a conically scanning
visual system analogous to
conically scanning radar
systems (Foster & Smyth, 1980).
The cyclic modulation of light
received by the eye as the cell
rotates provides a temporal
error signal to enable the cell to align its swimming path
with the source of light. It does this by steering, i.e.
controlling differentially the beating of the two flagella.
This tracking system enables the cell to swim more-or-less
directly in a helical path toward or away from light. Toward
or away is controlled by the phasing determined by whether
the cell responds to an increase or decrease in light level.
The monopolar directivity of the eye is absolutely critical
because it is the source of the temporal modulation as the
cell rotates. If the eye had uniform sensitivity in all
directions there would be no temporal modulation. The advantage of using a multilayer device compared to a single thick
absorptive layer one-quarter wave behind the pigment is il-
lustrated by the modulation function as a function of photon
energy (Fig. 2B). The position of the eye on the cell is
also precisely determined so that it is phase advanced with
respect to the differential control of the flagella. This
compensates for the delay required for signal processing
(Foster & Smyth, 1980). We hope with *Chlamydomonas* to understand how the geometry of its eye is specified.

INTRACELLULAR SIGNALING

In all cells, signals must be communicated from one place to another electrically and/or chemically. Simultaneous signaling must occur without crippling cross-talk between the various organelles of a cell. How does a cell do it? *Chlamydomonas* has a number of signaling pathways amenable to study. A single celled organism has the advantage over a multicellular one that its environment can be more precisely controlled. We have focused on how rhodopsin regulates phototaxis (Foster *et al.*, 1984) and gene expression (Foster *et al.*, 1988c). It is potentially also involved in circadian rhythm, its "down regulation" and its own degradation and synthesis. A variety of diseases are the result of faulty intracellular signaling. Cancers are failures of signaling from membrane receptors to control of cell growth (Darnell *et al.*, 1986). In cystic fibrosis the signaling from the membrane to control Cl⁻ channels is faulty (McPherson & Dorner, 1988). Retinitis pigmentosa are probably errors in the pathway of rhodopsin control of gene expression or in control of expression of opsin and other components of the photoreceptor (Foster *et al.*, 1988c).

Rhodopsin initiated behavior has been well studied in *Chlamydomonas* (Feinleib & Curry, 1971; Foster & Smyth, 1980; Nultsch 1983) and in vertebrate and invertebrate systems (Stryer, 1987). The ciliary photoreceptor cells of vertebrates appear to communicate their signal from rhodopsin to the next cell via a visual cascade involving three stages of amplification: GDP-GTP exchange on G-proteins, phosphodiesterase hydrolysis of cGMP, and closure of a cation channel. In rhabdomeric photoreceptor cell inositol trisphosphate is thought to be the second messanger. In *Chlamydomonas*, G-proteins, cGMP (P Hegemann, personal communication), and light activated channels are involved (JM Sullivan & KW Foster, unpubl.), but a complete description is not yet available. There appears to be only one stage of enzymatic amplification and hyperpolarization appears to occur. As with other photoreceptors the signal from the receptor is amplified, light adapted, signal compressed, and transmitted to the remote output structures (in *Chlamydomonas* its flagella). The signal processing functions are performed by light, voltage, ions (primarily Ca^{2+} and K^+) and probably chemically regulated ion channels in the plasmamembrane.

Using the cell-attached patch clamp technique to study

light-modulated current activity in *Chlamydomonas*, we observed: 1) regenerative currents probably mediated by Ca^{2+}
action potentials, 2) periodic macroscopic slow inward or
outward currents seen in continuous light and occasionally
associated with single-ion channel activity, 3) ON and OFF
responses of macroscopic current to blue-green (but not red)
light occurring in the region near the eye, and 4) blue-green
light modulated single channel activity (JM Sullivan & KW
Foster, unpubl.). The light-modulated currents were similar
to that found in ciliated photoreceptor systems. Hyperpolar-
izing plasmamembrane receptor potentials in the Chlamydomon-
ad, *Haematococcus pluvialis*, are graded with light and show a
rhodopsin-like action spectrum (Sineschekov *et al.*, 1976;
Litvin *et al.*, 1978). The behavioral output can be quanti-
fied on held cells by measuring the differential beating
response of the two cilia to light stimulation or the light-
dependent turning of free swimming cells (Hegemann & Marwan,
1988; RD Smyth & HC Berg, unpubl.). The accessibility of the
genetics and electrophysiology of *Chlamydomonas* offers an
opportunity to study this rhodopsin-mediated transduction
system in molecular detail.

In addition to regulating cell behavior, rhodopsin activation controls gene expression (Foster *et al.*, 1988c). We
demonstrated this with an assay for induction of retinal
synthesis based on the phototactic response of a blind mutant. In the dark the mutant fails to synthesize retinal,
but has opsin. When retinal synthesis is induced by light
the cells swim away from a source of light. Since the amount
of light required to trigger a phototactic response near
threshold is inversely proportional to the concentration of
rhodopsin, the decrease in amount of light necessary to generate that threshold response serves as a measure of the
amount of retinal synthesized in cells after induction. The
amount of light-induced retinal depends linearly on the light
exposure and on the rhodopsin concentration during the exposure. The action spectrum of light induction is identical
with that for phototaxis for which the receptor pigment is
rhodopsin and that incubation with all-*trans*-7,8-dihydrore-
tinal before light exposure shifts the action spectrum peak
for light induction 0.41 eV (-71 nm) (Fig. 3). The shifted
spectrum shows that this analog has been incorporated and is
active directly in the receptor. We concluded that the photopigment for induction of retinal synthesis is a rhodopsin.
This action spectrum technique is valuable for demonstrating
the identity of responsible photoreceptors and is generally

applicable (Foster *et al.*, 1988c; Smyth *et al.*, 1988).

Figure 3. Light induction action spectra following retinal (o) or 7,8-dihydro-retinal (•) incorporation into opsin.

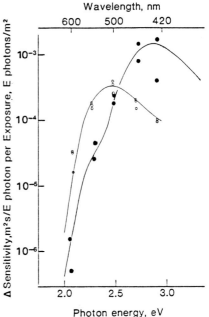

We measured the effects of transcription and translation inhibitors (at the effective dose determined by doing dose-response studies) added to the cell suspensions before, during and after light exposure. The light induction of retinal synthesis was inhibited at the transcription level at both choroplastic and nuclear sites, if the inhibitors were added before or during (but not af-ter) light exposure. These results and the time lag re-quired for induction of retinal synthesis suggest that gene expression is involved in the induction process and that both nuclear and chloroplast transcription are required for retinal synthesis in the light. The effects of the inhibi-tors on trisporic acid stimulation of retinal synthesis in the dark suggested that either nuclear or chloroplast trans-cription was sufficient (J Saranak, KW Foster, unpubl.). Using the conditions of light induction found from these studies we made a cDNA library selective for those mRNAs specifically enhanced by rhodopsin activation and test clones in order to find a molecular probe for the expressed genes which will allow us to trace their regulation (Petridou *et al.*, 1988).

In addition to the regulatory pathway of retinal synthe-sis, the enzyme, 15,15'-β-carotene dioxygenase, which cleaves β-carotene to retinal is feasible to study in *Chlamydomonas*. It is the enzyme in humans that makes dietary carotenes a source of vitamin A for vision and regulation of the repro-ductive system. In spite of this importance it is not well characterized. *Chlamydomonas* phototaxis provides an excel-

lent assay for its substrate. Carotene analogs are quickly
enzymatically converted to retinal analogs which are then
incorporated into opsin restoring phototaxis. With this
assay we are able to study the substrates and inhibitors of
the dioxygenase.

Finally, *Chlamydomonas* has the virtue of having many
sensory inputs which control gene expression including heat
shock, gamete induction, mating, flagella excision, and at
least three photopigments, blue light, rhodopsin and chlorophyll. Study of how these signals manage not to interfere
with each other promises to be most exciting.

RHODOPSIN ACTIVATION: A MODEL OF MEMBRANE RECEPTOR ACTIVATION

In rhodopsin, how does light absorption by N-retinylidene in rhodopsin's regulatory site causes rhodopsin's enzymatic site to promote GDP-GTP exchange and drive the visual
cascade? This signal communicates through the protein since
the regulatory site is in the interior of the protein and the
enzymatic site is on the cytoplasmic side. The activation
must also be transient and the protein made ready with the
optimum delay for another activation cycle. One particular
design has evolved which handles a wide range of regulators;
the homologous family of receptors includes the rhodopsin,
muscarinic, β-adrenergic, the oncogene *Hmas*, membrane cAMP,
substance K and serotonin receptors, and probably other membrane receptors such as the histamine, dopamine, opiate, and
prostaglandin receptors. Understanding their activation
mechanisms would be of enormous benefit in intervening in
control of their activity. One further would like to understand how these receptors are made selective, for rhodopsin
how does each rhodopsin absorb only a part of the visual
spectrum, and similarly how does a hormone receptor distin-
guishes between ligands.

Chlamydomonas has a bovine-like rhodopsin receptor (Foster *et al.*, 1984). The blind mutant which responded only to
light irradiances 1000x normal was restored to normal phototaxis sensitivity after addition of retinal. To prove that
this was due to the incorporation of retinals specifically to
opsin we incubated the blind mutant with retinal analogs
which shifted the spectrum and measured the threshold action
spectra (for details on techniques and analysis see Foster et
al., 1984, 1988c; Smyth et al., 1988; Hegemann et al., 1988).

All the analogs that we incubated with the cells restored
activity with approximately the same shifts as had been found
for bovine (*Bos*) rhodopsin and with an almost perfect fit to
the absorption curve of each pigment (Foster *et al.*, 1984;
Smyth *et al.*, 1988). This meant the added retinal formed a
chromophore with the receptor pigment in *Chlamydomonas* and
was a rhodopsin. Further it suggested that the environments
of the chromophore in the rhodopsins of *Chlamydomonas* and *Bos*
are nearly electrically equivalent.

Determination of Spectral Sensitivity of Rhodopsins

The primary determinant of what part of the visual spectrum a rhodopsin absorbs is the distribution of charge in the
neighborhood of the N-retinylidene chromophore (formed by
retinal making an imine bond with the ϵ-amino group of an
opsin lysine). We concluded this from the results of syste-
matically varying the charge distribution of the photoactiva-
ting chromophore by incorporating retinal analogs having
from two to seven conjugated double bonds, those with substi-
tuted groups of differing electronegativity, and the satura-
ted aldehyde n-hexanal. The opsin moieties in close proxim-
ity to the chromophore also influence the charge distribution
of each chromophore and such influence appears as shifted
rhodopsin spectra. The response to incorporation of the
acyclic retinal analog series (Fig. 4) are in the normal
range of sensitivities implying normal incorporation and
quantum efficiency. With different chromophores and corres-
pondingly different charge distributions, we observed a sys-
tematic variation in the action spectra maxima which enabled
us to derive a general model for spectral sensitivity of
rhodopsins (Foster *et al.*, 1988b).

The absorption of one photon causes excitation of one
electron in rhodopsins and the transition energy of a one-
electron process is the difference between the energy levels
of the initial and final state. This transition energy
equals the energy of the absorbed photon and corresponds to
the position of the absorption band in the spectrum. The
transition energy may be changed by perturbations of either
the initial or final state, for instance by presence of an
external charge in the vicinity of the chromophore. To
understand the influence of an external charge on the one
electron transition the concept of effective charge is useful. Any state of an electron can be perturbed by the

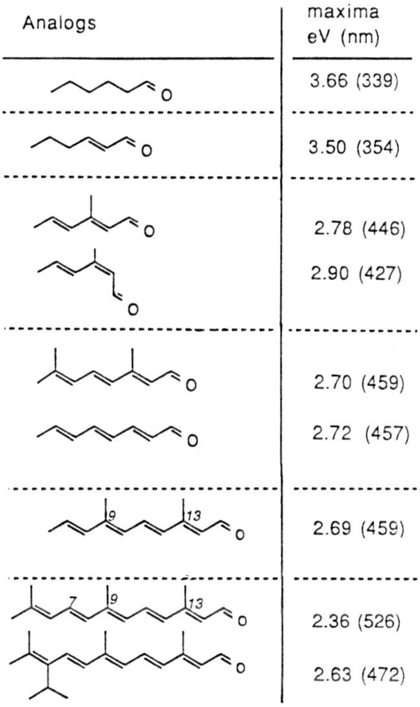

Analogs	maxima eV (nm)
Figure 4. Acyclic retinal analogs having from 1 to 6 double bonds.	
	3.66 (339)
	3.50 (354)
	2.78 (446)
	2.90 (427)
	2.70 (459)
	2.72 (457)
	2.69 (459)
	2.36 (526)
	2.63 (472)

Figure 4. Acyclic retinal analogs having from 1 to 6 double bonds.

presence or absence of other electrons (screening) as well as by the presence of an external charge. If there existed only an ion and a chromophore together then the effect of the perturbing charge on the chromophore would be maximum. However, screening of charge by other molecules reduces the effect, which consequently reduces the effective charge of the ion.

An energy level diagram (Fig. 5A) summarizes the consequences of interaction of effective charges with an electron in the initial (ground) state and the final (excited) state. As a reference level, we use E_{vacuum}, the energy that the electron would have if it were completely removed from the chromophore and had zero kinetic energy. E_{vacuum} minus the energy of the electron state is the binding energy of the electron. A *positive* charge (provided by a proton, electronegative substituent, external charge group, or ion) in effect *removes* a small negative charge from the chromophore, thus increasing the binding energy of the electron that will be excited by the photon and lowering its energy level. Conversely, an external *negative* charge acts as a source of electrons and *adds* negative charge, thus reducing the binding energy and raising the energy level of the bound electron state.

An external charge or substituent on the chromophore affects both the ground and excited states. The effect on the transition energy depends on the relative magnitude of the two effects. The effective charge has its greatest effect on the state whose electrons are closer. In retinyl imines the electron distribution shifts toward the CN end of the molecule upon absorption of a photon. Therefore a charge located toward the CN end has a greater effect on the excited

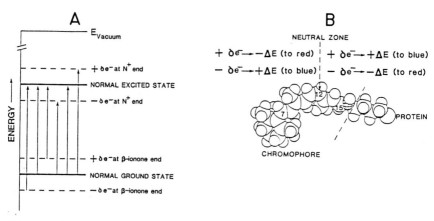

Figure 5. Energy level diagram (A) and space filling model of the N-retinylidene showing the neutral zone and the effect of external charges (B). The hatched atom is N.

state; a charge located toward the free end has a greater effect on the initial state of the photoexcited electron. Maximal differential effect occurs when the perturbing charge is at either end of the chromophore. Between the two ends is a neutral zone where an external charge has equal effect on the initial and final states and therefore does not change the transition energy. For N-retinylidene in *Chlamydomonas*, *Bos* and in solution the neutral zone was determined to be at position C-12 as depicted in Fig. 5B.

As a consequence protonation of the imine N causes a large red shift (decrease in the transition energy). Additional shift is caused by reduced screening in the nonpolar opsin environment as compared to retinal in a polar solvent like methanol and by a negative charge in the binding site toward the ring end from the neutral zone. The evidence for this is the sharp discontinuity of the shift between three and four double bonds as shown in Fig. 6. We used the analogs as electronic gauges laid in the binding site with one end fixed at the CN bond. When the charge is at the free end of the chromophore it has maximal differential effect in its influence of the ground and excited states and maximally raises the energy of the ground state relative to the excited state and thus maximally lowers the transition energy. For *Chlamydomonas* this locates the charge near C-10 or C-11 (Fig. 5B) and most importantly is consistent with the effect of synthetic analogs in which charges have been fixed at known positions (Baasov & Sheves, 1985). The same conclusions have been confirmed by substitution of strongly electronegative groups (small effective positive charges) along the chromophore (J Saranak, unpubl.).

Figure 6. The difference in spectral maxima between protonated N-retinylamine in methanol and in *Chlamydomonas* as a function of the number of double bonds of the retinal analog.

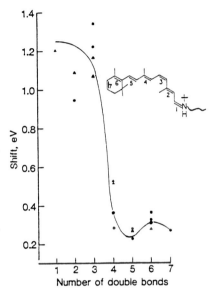

The spectral sensitivity of rhodopsins can then be explained by the perturbations of the excitation of retinal by screening of the initial and/or final state of the electron that is photoexcited. The perturbations are introduced in the form of the effective charges of amino acids in the protein, attachment of different electronegative groups to the retinal, or by changes in the effective charge of part of the chromophore. The state they influence is determined by their relative positions with respect to the chromophore. Using these principles one can explain human red-green color blindness very simply. The inability to distinguish red from green results from the red and green pigments being too similar in their absorption spectra. Their differences are due to the number and placement of hydroxyls in the retinal binding site of their respective opsins (Foster *et al.*, 1988b). Long ago it was observed that the absorption maxima of visual pigments in fish (Dartnall & Lythgoe, 1965), monkeys and humans (Bowmaker & Dartnall, 1980) were not scattered randomly across the spectrum but occurred in discrete clusters, 0.028 eV apart. The quantal nature is expected from the limited number of polar changes that are available from amino acid substitutions in the retinal binding site. The difference between human red (2.20 eV, 564 nm) and green (2.32 eV, 534 nm) corresponds to 4 hydroxyls, possibly as drawn on the structures in Fig. 7. Loss of any of these hydroxyls in the protein as a result of altered DNA code due to recombination of the red

and green genes (Nathans et al., 1986) will therefore result
in color blindness. In a similar manner alteration of amino
acids in the binding or regulatory sites of hormone receptors
will change the selectivity of the receptors to different
potential ligands.

Figure 7. Tentative
schematic arrangement
of protein helices
(circles) of two rho-
dopsins viewed from the
exterior. Green cone
opsin has two OH's on
helix 5; red cone one
OH on helices 1 and 6.
The handedness of the
protein has been deter-

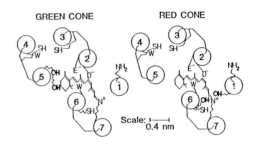

GREEN CONE   RED CONE

Scale: 0.4 nm

mined by analysis of the spectral shifts due to the two car-
boxyl groups in red and green pigments compared to the *Bos*
rod and *Chlamydomonas* pigment.

Mechanism of Rhodopsin Activation

In the first step of vision, a photon is absorbed by
retinal, which lies in the regulatory site of rhodopsin.
Upon photoexcitation, charge is redistributed as the electron
density shifts toward the imine (C=N) end of retinal. The
charge redistribution triggers double-bond isomerization of
retinal, as well as bleaching (release of retinal from its
binding site), changes in the apparent pK_a of the imine N,
proton motion in opsin and potentially direct excitation of
rhodopsin. *Post hoc* reasoning lead to reasonable hypotheses
that one or another of these events activated the visual
cascade with *cis-trans* isomerization (Kropf & Hubbard, 1958)
becoming the most popular. However, no clear evidence for or
against any particular hypothesis was found. This was because absorption spectroscopic techniques monitor retinal in
the regulatory site in the middle of rhodopsin. These techniques, by themselves, cannot determine the state of activation of the enzymatic site on the cytoplasmic surface of
rhodopsin. For this purpose one must have a measure of the
actual activation, such as an enzyme assay for products in
the visual cascade or an *in vivo* assay such as phototaxis in
Chlamydomonas. The *in vivo* assay has proved to be orders of
magnitude more sensitive and easier to carry out than the

present enzyme assays allowing a wider range of analogs to be tested more reliably.

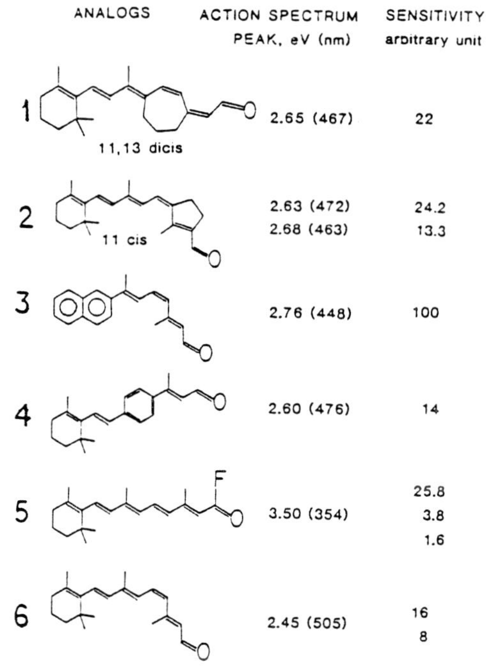

Figure 8. Retinal analogs: 1-4 are isomerization locked at particular double bonds, 5 forms an amide bond with the lysine N, and 6 is probably the normal chromophore, 11-*cis*-retinal.

ANALOGS	ACTION SPECTRUM PEAK, eV (nm)	SENSITIVITY arbitrary unit
1 11,13 dicis	2.65 (467)	22
2 11 cis	2.63 (472) 2.68 (463)	24.2 13.3
3	2.76 (448)	100
4	2.60 (476)	14
5	3.50 (354)	25.8 3.8 1.6
6	2.45 (505)	16 8

We have tested whether isomerization of retinal is required to activate the sensory function of rhodopsin by measuring the activity of retinal analogs that were blocked from complete isomerization (Fig. 8). These analogs twist transiently (< 100 ps) before returning to their original shape. We wondered whether their comparative geometrical rigidity and short photocycle would inhibit response. We also reduced the chromophore to the minimum necessary for efficient activation, i.e. n-hexanal (Fig. 4), and tested whether protonation of the N or release of retinal from its binding site was required. Our results lead to the following conclusions (Foster *et al.*, 1988a) with respect to the rhodopsin in *Chlamydomonas*: 1) Blocking of individual double-bond isomerizations does not prevent the activity of rhodopsin. Therefore, if isomerization is required at all, it must not matter at what position along the chromophore that it occurs. This conclusion is further supported by the normal activity of n-hexanal which has no C=C bonds to be isomerized. Its normal efficiency further implies that the bulk of retinal is not essential. 2) The positive activity of acid fluoride retinal (5 in Fig. 8) which forms an amide bond with the lysine N and therefore is not bleached or protonated at the N, shows that neither bleaching nor protonation is required for the activity of rhodopsin.

3) Changes in molecular geometry of the chromophore may be unnecessary for protein activation. 4) Incorporation of the 7-membered analog (1 in Fig. 8) into bovine rhodopsin blocked bathorhodopsin formation and subsequent bleaching. However, transiently excited species were present for 100 ps (Buchert et al., 1983). Its activity in Chlamydomonas implies that changes in the protein must be triggered within the 100 ps period before the chromophore returns to its original ground state.

Our results suggest that the rhodopsin of Chlamydomonas is activated as a direct result of the charge redistribution in the excited state of the chromophore and not via isomerization. As described, spectral sensitivity is controlled by the protein surrounding the chromophore. Therefore, it is reasonable that the large change in the distribution of electron density on excitation of the chromophore (Mathies & Stryer, 1976) switches the state of the protein to the enzymatically active conformation. A number of theoretical papers (Salem & Bruckman, 1975; Lewis, 1978; Warshel, 1978) have previously suggested electronic activation mechanisms. Our results suggest that there are two photocycles to be distinguished, that of the chromophore and that of opsin and two distinct activities of opsin, namely activation of the enzymatic site and recycling its chromophore. More details of the arguments may be found in Foster et al. (1988b).

The current status is that: 1) a large steric change such as cis-trans isomerization about C-7, C-9 or C-11 (but not C-13) may be required for bleaching but not for activation; 2) rotation about the C=N bond and/or changes in the pKa of the N conceivably could play a role in activation but not in bleaching; and 3) direct electronic excitation of the protein is the best candidate for activating the protein, but may not be required for bleaching. Because several effects triggered by the charge redistribution within the chromophore may be involved in different aspects of visual signaling, the charge redistribution should be considered the primary step critical in visual transduction. Homology of the various eukaryotic rhodopsins (Martin et al. 1986; K. Volpp et al. unpub.) suggests that the activation mechanism supported by the data on the rhodopsin of Chlamydomonas is probably the mechanism for all other eukaryotic rhodopsins. Since rhodopsin is a member of a membrane receptor family we can anticipate determining how these GDP-GTP exchange enzymes are activated and how they evolved from one to another.

REFERENCES

Baasov T, Sheves M (1985). Model compounds for the study of spectroscopic properties of visual pigments and bacteriorhodopsin. J Am Chem Soc 107:7524-7533.

Bowmaker JK, Dartnall HJA (1980). Visual pigments of rods and cones in a human retina. J Physiol 298:501-511.

Buchert J, Stefancic V, Doukas AG, Alfano RR, Callender RH, Pande J, Akita H, Balogh-Nair V, Nakanishi K (1983). Picosecond kinetic absorption and fluorescence studies of bovine rhodopsin with a fixed 1-ene. Biophys J 43:279-283.

Darnell J, Lodish H, Baltimore D (1986). "Molecular Cell Biology." New York: Scientific American Books, pp 1059-1060.

Dartnall HJA, Lythgoe JN (1965). The spectral clustering of visual pigments. Vision Res. 5:81-100.

di Francia GT (1949). Retina cones as dielectric antennas. J Opt Soc Amer 39:324-325.

Ehrenberg CG (1838). "Die Infusionsthierchen als vollkonnene Organismen." Leipzig: Voss, pp. 547.

Feinleib ME, Curry GM (1971). The nature of the photoreceptor in phototaxis. In Loewenstein WR (ed): "Principles of Receptor Physiology," New York: Springer-Verglag, pp 366-395.

Foster KW, Saranak J, Derguini F, Zarrilli GR, Johnson R, Okabe M, Nakanishi K (1988a). Activation of *Chlamydomonas* rhodopsin *in vivo* does not require isomerization of retinal. Biochemistry (in press).

Foster KW, Saranak J, Dowben PA (1988b). Color sensitivity and activation of rhodopsin: effects of retinal analogs on *Chlamydomonas* behavior. Submitted May 1988 to Biophys J.

Foster KW, Saranak J, Patel N, Zarrilli G, Okabe M, Kline T, Nakanishi K (1984). A rhodopsin is the functional photoreceptor for phototaxis in the unicellular eukaryote *Chlamydomonas*. Nature 311:756-759.

Foster KW, Saranak J, Zarrilli GR (1988c). Autoregulation of rhodopsin synthesis in *Chlamydomonas reinhardtii*. Proc Natl Acad Sci USA 85:6379-6383.

Foster KW, Smyth RD (1980). Light antennas in phototactic algae. Microbiol Rev 44:572-630.

Harold FM (1986). "The Vital Force: A Study of Bioenergetics." New York: Freeman, pp 524-568.

Hegemann P, Hegemann U, Foster KW (1988). Reversible bleaching of *Chlamydomonas reinhardtii* rhodopsin *in vitro*. Photochem Photobiol 48:123-128.

Hegemann P, Marwan W (1988). Single photons are sufficient

to trigger movement responses in *Chlamydomonas reinhardtii* Photochem Photobiol 48:96-106.

Kropf A, Hubbard R (1958). The mechanism of bleaching rhodopsin. Ann Rev NY Acad Sci 74:266-280.

Lewis A (1978). The molecular mechanism of excitation in visual transduction and bacteriorhodopsin. Proc Natl Acad Sci USA 75:549-553.

Litvin FF, Sineshchekov OA, Sineshchekov VA (1978). Photoreceptor electric potentials in the alga *Haematococcus pluvialis*. Nature 271:476-478.

Martin RL, Wood C, Baehr W, Applebury, ML (1986). Visual pigment homologies revealed by DNA hybridization. Science 232:1266-1269.

Mathies R, Stryer L (1976). Retinal has a highly dipolar vertically excited singlet state: implications for vision. Proc Natl Acad Sci USA 73:2169-2173.

McPherson MA, Dorner RC (1988). Cystic fibrosis: a defect in stimulus-response coupling. Trends Biochem Sci 13:16-18.

Nathans J, Thomas D, Hogness DS (1986). Molecular genetics of human color vision: The genes encoding blue, green, and red pigments. Science 232:193-202.

Nultsch W (1983). The photocontrol of movement in *Chlamydomonas*. Symp Soc Exp Biol 36:521-539.

Petridou S, Kindle K, Saranak J, Foster KW (1988). Rhodopsin regulation of gene expression in *Chlamydomonas reinhardtii*. Proc. Yamada Science Conference XIII, Kyoto Japan.

Poole RK, Trinci, APJ (1987). "Spatial Organization in Eukaryotic Microbes." Washington DC, IRL Press, 1-140.

Salem L, Bruckmann P. (1975). Conversion of a photon to an electrical signal by sudden polarization in the N-retinylidene visual chromophore. Nature 258:526-528.

Sineshchekov OA, Andrianov VK, Kurella GA, Litvin FF (1976). Bioelectric phenomena in unicellular flagellar alge and thier connection with phototaxis and photosynthesis. Fiziologiya Rastenii 23:229-237.

Smyth RD, Saranak J, Foster KW (1988). Algal visual systems and their photoreceptor pigments. Prog Phycol Res 6:(in press).

Sogin ML, Elwood HJ, Gunderson JH (1986). Evolutionary diversity of eukaryotic small-subunit rRNA genes. Proc Natl Acad Sci USA 83:1387-1393.

Stryer L (1987). Visual transduction: design and recurring motifs. Chemica Scripta 27B:161-171.

Warshel A (1978). Charge stabilization mechanism in the visual and purple membrane pigments. Proc Natl Acad USA 75:2558-2562.

Section IV.

Model Systems for Photoreactions and Rhythms

Algae as Experimental Systems pages 233–248
© 1989 Alan R. Liss, Inc.

Algae as Model Systems in the Study of Photosynthetic Membrane Organization

Kenneth R. Miller and Laurel Spear-Bernstein

Division of Biology & Medicine, (K. R. M.), Brown University, Providence, Rhode Island 02912, and Department of Molecular Biology, (L. S.-B.) Scripps Clinic and Research Foundation, La Jolla, California 92037

The photosynthetic membrane is the most widespread energy-transducing structure in nature. The architecture of this membrane is intimately related to its role in capturing the energy of sunlight and converting that energy to usable form of chemical energy. Photosynthetic membranes are present in plants, algae, and photosynthetic bacteria. The similarities and differences which exist among these groups of organisms provide an opportunity for comparative studies to address some of the ourstanding problems of photosynthetic membrane organization.

The photosynthetic membrane is a closed sac, whose osmotic integrity allows the development of a potential gradient which drives the synthesis of ATP. The membrane contains four major classes of functional complexes. 1) A light-harvesting system, composed of pigment-protein complexes: this system absorbs sunlight and channels the excitation energy towards a photochemical trap where it can be used for energy transduction. 2) Photosynthetic reaction centers, which contain both pigment and protein,and provide an environment in which excitation energy can promote a charge separation event. Plants and algae have two different types of reaction centers, which function in series, while photosynthetic bacteria have only a single type of reaction center. 3) An electron-transport system, which passes electrons by means of a series of redox reactions: in oxygen-evolving photosynthetic organisms, one of the components of the electron transport system is involved with the removal of electrons from water, and the release of oxygen as a byproduct of photosynthesis. 4) An ATP synthesizing system. The ATP-synthetases of this system are closely related to the similar enzymes in mitochondria and bacteria, and they are involved in the synthesis of ATP in response to a proton gradient which develops across the membrane as a consequence of electron transport.

PHOTOSYNTHETIC MEMBRANES IN HIGHER PLANTS

Most studies on photosynthetic membranes have centered around the organization of the thylakoid membranes in the chloroplasts of plants and green algae (see Murphy, 1986). Thylakoid membranes consist of an extensive network of saclike membranes which are appressed, or stacked, at their outer surfaces. In plants, these large, regular stacks may reach heights of 30-40 thylakoids and are known as "grana". The phenomenon of membrane stacking is widespread, although not universal, among photosynthetic organisms, and its importance is one of the major questions which the research described here has sought to approach. One of the major structural tools with which we have sought to investigate thylakoid organization is the freeze-etch technique. This preparative procedure for electron microscopy allows a single biological membrane to be examined from four viewpoints. In freeze-fractured samples, the membrane splits along a central plane, exposing two complementary fracture faces which reveal the internal organization of the membrane, and in deep-etched samples, the sublimation of frozen buffer from the membrane allows the examination of the true inner and outer surfaces of the membrane (see Miller, 1981). The appearance of a thylakoid membrane in a freeze-fractured preparation is shown in Figure 1.

There are four distinct fracture faces visible in a freeze-fractured preparation (Fig. 1). The origin of these four fracture faces was explained in a study of thylakoid membranes isolated from an algal system. Goodenough and Staehelin (1971) showed that only two fracture faces were visible in a mutant line of *Chlamydomonas* which lacked membrane stacking. Inferring that thylakoid stacking might be responsible for the two additional fracture faces, they carried out experiments in which thylakoid membranes were artificially unstacked and restacked by changes in ionic concentration. These experiments established quite clearly that stacked and nonstacked membranes have different internal structures. This accounted for the four fracture faces: two are derived from membrane splitting in stacked regions (PFu and EFu, which refer to "Protoplasmic Face, unstacked" and "Ectoplasmic Face, unstacked," respectively) and two are derived from membrane splitting in nonstacked regions (EFs and PFs, which refer to the corresponding fracture faces in stacked membranes). The labeling protocol for freeze-fractured membranes was described in Branton *et al.* (1975).

Freeze-fracture studies provided one of the early hints that thylakoid membrane organization might be different in stacked and unstacked regions of the thylakoid membrane system. Other studies (reviewed in Anderson and Andersson, 1982) showed that the differences between stacked and nonstacked regions of the thylakoid membrane were profound: the nonstacked regions had a high chlorophyll a/b ratio, as well as nearly all of the photosystem I (PS I) units and ATP-synthetase complexes; the stacked regions had a low chlorophyll a/b ratio, as well as nearly all of the photosystem II (PS II) units. These differences were summarized by Anderson and Andersson (1982) who coined the term "lateral heterogeneity" to describe the local differences along the length of the thylakoid membrane.

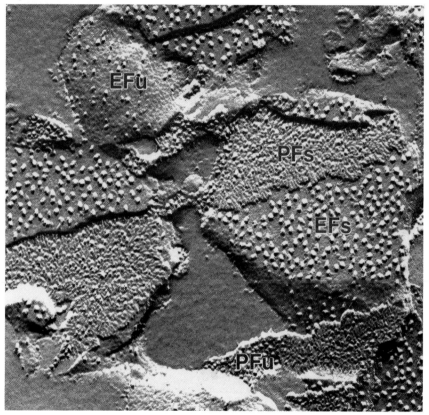

Figure 1: Freeze-fracture image of spinach thylakoid membranes. Four fracture faces are labeled, two derived from stacked membranes (PFs and EFs) and two derived from nonstacked membranes (PFu and EFu). Magnification: 125,000.

REASONS FOR WORKING WITH ALGAL SYSTEMS

The organization of thylakoid membranes has a biochemical component as well as a structural one, and an accumulating literature (reviewed by Murphy, 1986) has determined the major polypeptide and pigment components of the photosynthetic membrane, described their molecular weights and in a few cases clarified their transbilayer organization (Sayre, 1987). Although much of this work has been carried out in higher plants, there is a series of examples in which algal systems have been crucial in advancing our understanding of the thylakoid. There are two general reasons for the use of algae as experimental systems in photosynthesis, 1) Many algal systems enjoy the advantage of simplicity over higher plants. The use of such systems allows the investigator to focus in on a narrower range of variables and to design experiments in a less complicated system. 2) The diversity of algae, particularly if the cyanobacteria (blue-green algae) and photosynthetic bacteria are included, provides a series of photosynthetic systems with such a wide range of structures and adaptations that many questions can be answered by means of a comparative approach.

One of the classic examples of the fruits of comparative studies was given by van Niel (1935), who noted that the sulfur photosynthetic bacteria did not release oxygen in photosynthesis, but instead produced elemental sulfur from hydrogen sulfide (H_2S). Correctly reasoning that H_2S was playing the role of electron donor in bacterial photosynthesis, he concluded that water (H_2O) was the equivalent electron donor for green plants, and therefore photogenic oxygen release came not from carbon dioxide (as many investigators thought), but from water.

DETAILED STRUCTURAL STUDIES ON THYLAKOID MEMBRANES

Work in our laboratory has taken two general directions, each of which has involved the use of unicellular photosynthetic organisms as model system. One direction, which involves the comparative approach, will be discussed shortly. Our other approach to the study of the photosynthetic membrane has been to develop systems which can be visualized at high resolution. A key tool in this respect is the application of Fourier imaging techniques to the study of membrane structure. These techniques have a crucial requirement: the need for a highly ordered structure in which the images of hundreds or thousands of individual subunits may be averaged together to obtain a noise-free image with excellent structural detail. In some instances, experimental manipulation causes the higher plant thylakoid to assume an ordered structure which is suitable for Fourier analysis.

The photosynthetic membrane's inner surface, as seen by means of the deep-etching technique (Figure 2), shows stacked regions characterized by tightly packed tetramer-like structures, which have been associated with the PS II complex (Miller, 1981). In some preparations, these tetramers are organized into regular arrays, enabling the use of Fourier averaging to prepare a filtered image showing the detailed structure of a few tetramers. Although the actual resolution of the image is not improved by the processing technique, the reduction in noise allows a clear view of the tetramer structure, which Seibert *et al*.(1987) have associated with elements of the oxygen-evolving apparatus and PS II. The rarity of such regular regions has limited their usefulness in research on thylakoid structure.

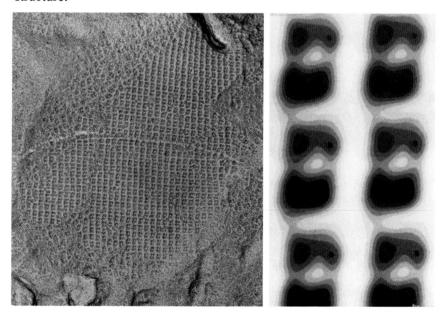

Figure 2: The inner surface of the thylakoid membranes of many groups, including green algae and plants, is characterized by a particle which has been associated with the photosystem II reaction complex. The left side of the figure shows a deep-etched surface view of a regular lattice of photosystem II particles from a spinach thylakoid membrane. Magnification: 100,000. The right side of the figure shows a Fourier-filtered image of the structure, clarifying its tetrameric organization. *(Courtesy of Jenny Hinshaw)*.

However, *Rhodopseudomonas viridis*, a photosynthetic bacterium, provides a model system in which the basic structure of the thylakoid membrane can be studied by means of Fourier analysis. The regularity with which subunits are arranged in this thylakoid has allowed us to prepare two-dimensional (Miller, 1979) and three-dimensional (Miller, 1982) maps of this membrane. The interpretation of these maps is made far easier by the availability of sheetlike two dimensional crystals (Miller, 1987) and particularly by the extraordinary three dimensional crystals of the *R. viridis* reaction center analyzed by Diesenhofer, Michel and their associates (Diesenhofer *et al.*, 1985). These studies led to an atomic level map of the four polypeptides and associated pigments in the reaction center, and established the idea that the reaction center was surrounded by a ring of light-harvesting pigment-protein complexes. More recently, Boekema *et al.* (1987) have show the application of other image analysis techniques to photosystem I complexes isolated from *Synechococcus*.

Studies on these and other systems will continue to yield valuable information on thylakoid membranes, and this is particularly true in cases where detailed genetic and molecular manipulations of membrane polypeptides have become possible (Gantt, this volume; Grossman *et al.*, this volume)

REGULATION OF EXCITATION ENERGY DISTRIBUTION

Although photosynthetic systems share a number of general similarities, one of the areas in which they show the greatest diversity is the mechanisms by which they harvest and distribute energy to their photosystems. It is in this area that the use of algal systems has been particularly useful, and this will clearly continue to be true. One of the most important developments in photosynthesis research in the past several years has been the elucidation of the mechanism by which excitation energy is distributed between PS I and PS II in higher plants. Because the wavelength of illumination which drives photosynthesis varies during the course of a day, it is important to maintain flexibility in the manner in which light energy is distributed between the two photosystems. Recent work has shown that changes in the pattern of excitation energy distribution in green plants center arounds a chlorophyll-protein complex known as light harvesting complex II (LHC-II) (see Staehelin & Arntzen, 1983,).

In brief, the model which has developed (Fig. 3) shows that LHC-II can be phosphorylated by a membrane-bound protein kinase which is sensitive to the redox state of plastoquinone (or possibly cytochrome B_6/f (Wollman &

Lemaire, 1988)), one of the electron carriers between PS I and PS II. When PS II is over-excited, the kinase is activated and a fraction of the LHC-II molecules are phosphorylated. The addition of the bulky charged phosphate groups causes these LHC-II molecules to migrate from the stacked regions of the thylakoid system into the nonstacked regions, resulting in a decrease in the light-harvesting pool available to PS II and an increase in the pool available to PS I. The system is self-adjusting, and over-excitation of PS I has the opposite effect, inhibiting the kinase and causing a reverse migration of LHC-II back into the stacked regions where PS II is found.

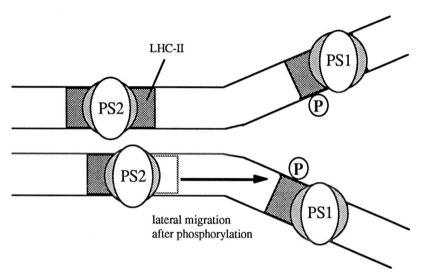

Figure 3: A model for the role of LHC-II in the distribution of excitation energy between the photosystems. When photosystem II is over-excited, a protein kinase phosphorylates some of the LHC-II pool, causing phosphorylated molecules to migrate to the nonstacked regions of the membrane, where they can interact with photosystem I. The reverse process occurs when photosystem I is over-excited. [drawing based on data described in Staehelin & Arntzen, 1983].

Although this model elegantly explains how excitation energy might be divided between the two photosystems, it does not answer a key question: why do thylakoid membranes stack? Although there is clear evidence that LHC-II mediates membrane stacking (Allen, 1983), the notion that membrane stacking exists in order to make it possible for LHC-II to regulate energy distribution is not tenable (see Miller & Lyon, 1985). The key piece of evidence in this regard is the fact that red algae, whose photosynthetic membranes do not stack, are efficient regulators of energy distribution between the photosystems. As noted by Biggins and coworkers, these algae also lack a phosphoprotein regulatory system (Biggins *et al.*, 1984; Biggins & Bruce, 1987).

Algal Patterns of Energy Distribution

All known chlorophyll a containing organisms possess two photosystems. The light-harvesting systems of these species may be grouped into two categories: 1) those which are completely contained within membrane-bound pigment-protein complexes; and 2) those which include phycobiliproteins bound at the surface of the photosynthetic membrane. The latter group includes cyanobacteria as well as the red and cryptophyte algae, whereas the former includes all other algae and plants. With the exception of the cryptophytes, phycobiliprotein-containing organisms do not have stacked thylakoids. Organisms which lack phycobiliproteins do have stacked thylakoids, although not all workers agree on this point.

As elegantly explored in this volume by Gantt and her coworkers (Gantt, this volume), the phycobilisome is an extramembrane assembly of pigment-protein complex which resides on the outer surface of the thylakoid membrane. Phycobilisomes are most closely associated with PS II, and in some species a direct correlation has been made between phycobilisomes and intramembrane particles associated with PS II (Giddings *et al.*, 1983). Energy harvested by phycobilisomes is preferentially directed to PS II, although this can be adjusted by means of "state transitions" which accomplish an even distribution of energy between PS I and PS II. In contrast to the situation in higher plants, there is no evidence for a lateral segregation of PS I and PS II, no membrane appression, and no evidence for a protein phosphorylation mechanism (Biggins & Bruce, 1987). The evidence from cyanobacteria and red algae leads to the conclusion that lateral heretogeneity and membrane stacking are not required for the efficient distribution of excitation energy between the photosystems.

Does the presence of phycobiliproteins make it impossible for thylakoid membranes to stack? The cryptophyte algae contain phycobiliproteins as well as internal light-harvesting complexes which contain chlorophylls a and c_2 (Ingram & Hiller, 1983). The structural organization of the cryptophyte thylakoid membrane in freeze-fracture (Fig. 4) is remarkably similar to thylakoid of plants and green algae, and displays a classic lateral heterogeneity of components on the two fracture faces (Spear-Bernstein & Miller, 1985; Duarte & Vesk 1983). As first suggested by Gantt *et al.*. (1971) and confirmed by immunolabeling (Spear-Bernstein & Miller,

1987), the phycobiliproteins in cryptophytes are located within the lumen of the thylakoid sac (Fig. 5) These light-harvesting pigments do not form distinct phycobilisomes, and their location within the thylakoid sac leaves the outer surfaces of the membrane free to form stacks.

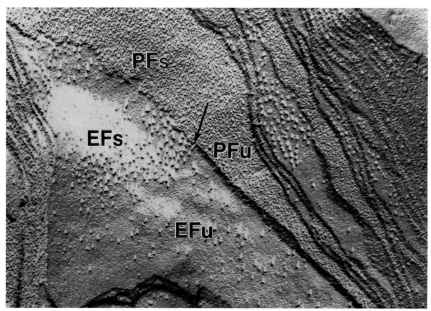

Figure 4: Freeze-fractured thylakoid membranes from the cryptophyte alga *Rhodomonas lens*. The internal structure of these photosynthetic membranes is remarkably similar to that of green algae, despite the presence of phycobiliproteins in the light-harvesting apparatus. The concentration of large particles in the EFs fracture face is clear evidence for a lateral segregation of components in this system, and illustrates "true" stacking according to our criteria. The arrow indicates a point of contact between two thylakoids.   Magnification: 100,000.

The question is sometimes raised as to whether the appression which is observed in cryptophytes and similar species is "true stacking," or merely a close association of membranes which have no specific interaction between them. Our view on this matter is that "true stacking" should involve an internal reorganization of the thylakoid membrane which can be recognized at the level of freeze-fracturing. Because cryptophyte thylakoids display four distinct fracture faces (Spear-Bernstein & Miller, 1985) as well as a distinct lateral segregation of components in stacked regions (Fig. 4), the thylakoid membranes of these organisms can truly be classified as stacked.

Figure 5: Immunogold labeling of phycoerythrin on thin sections of *Rhodomonas lens* . The concentration of label over the lumen of the thylakoid suggests that this pigment is contained within the thylakoid as first suggested by Gantt *et al* (1971). Magnification: 100,000.

The cryptophyte algae provide a clear demonstration of the fact that membrane stacking and phycobiliprotein-based light harvesting may occur in the same organism. This is an important insight which we may use in assessing the importance of membrane stacking in plants. However, until quite recently one of the difficulties with extrapolating algal research to higher plants was the fact that there were no plausible prokaryotic ancestors for the eukaryotic chloroplast: the light-harvesting and primary pigment systems of the cyanobacteria suggested an ancestry for the red algae, but not for the greens or for higher plants. In 1975 this situation was changed by the discovery of a prokaryotic organism which contained chlorophylls a and b (Lewin, 1975). *Prochloron* (as it is now known) was the basis for some initial studies of membrane organization (Giddings *et al.*, 1980), but the difficulty of culturing this organism in the laboratory limited its usefulness as an experimental system. The discovery of a second prochlorophyte organism, *Prochlorothrix hollandica* , (Burger-Wiersma *et al.*, 1986) which could be cultured easily made it possible to conduct more extensive investigations of prochlorophyte membrane organization.

The photosynthetic membranes of *Prochlorothrix hollandica* are organized into distinct stacks (Fig. 6), and its appearance in freeze-fracture leaves no doubt that membrane stacking occurs in this prochlorophyte. Four fracture faces are observed, two derived from membrane splitting in stacked regions and two from membrane splitting in non-stacked regions, and the overall pattern of fracture faces is remarkably similar to that observed in a higher plant thylakoid (cf. Figs. 1 and 7).

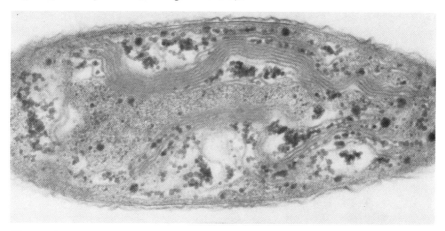

Figure 6: Thin section of the chlorophyll b -containing prochlorophyte *Prochlorothrix hollandica* . The photosynthetic membranes of this organism appear in stacks of 2 to 5 thylakoids. Magnification: 75,000.

Studies on the membrane surfaces of this organism revealed another fundamental similarity to higher plants. The characteristic tetramer structure, associated with the oxidizing side of PS II on the inner surface on the thylakoid membrane, is observed in *Prochlorothrix* as well as in higher plants (Fig. 8). In addition, the concentration of these tetramers into stacked regions parallels the concentration of EFs fracture faces particles into stacked regions, leading to the tentative conclusions that, as in higher plants, these tetramers are the structural equivalents of the PS II complex. *Prochlorothrix*, therefore, has a thylakoid structural organization which is very similar to that of higher plants. Interestingly, as Bullerjahn *et al.* (1987) have shown, the protein composition of the *Prochlorothrix* thylakoid is markedly different from that of higher plants.

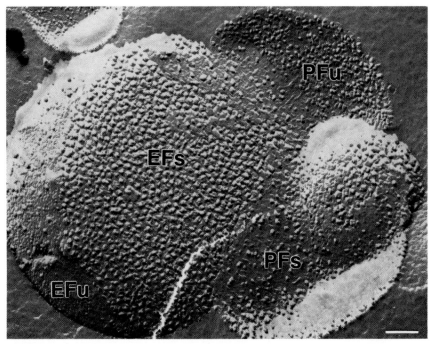

Figure 7: Freeze-fracture image of *Prochlorothrix hollandica* thylakoid membranes. As in green algae and plants, four fracture faces are visible, and there is a distinct lateral segregation of components into stacked regions of the membrane. Magnification: 125,000.

Finally, studies on prochlorophytes (Schuster *et al.* (1984) Snyder & Biggins, personal communication) and cryptophytes (Snyder & Biggins, 1987) have thus far failed to establish the existence of a true state transition (energy-regulating mechanism) in either class of organism. Thus, these organisms may present us with examples of thylakoid membranes in which lateral heterogeneity and thylakoid stacking are not associated with a mechanism of excitation energy distribution.

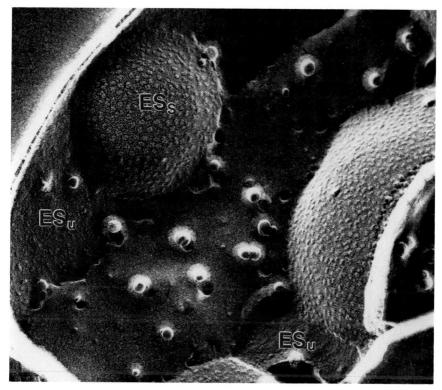

Figure 8: The inner surface of *Prochlorothrix hollandica* , revealed by deep-etching, displays tetramers which are remarkably similar to those found in green algae and higher plants. The occurrence of this structure in these prokaryotes provides some support for the suggestion that prochlorophytes may be related to the evolutionary ancestors of green algal chloroplasts. Magnification: 100,000.

IMPORTANCE OF THYLAKOID STACKING: LESSONS FROM THE ALGAE

Although considerable information on the mechanism of thylakoid stacking has emerged from studies on higher plants and green algae, these studies have not yet developed a clear rationale for the widespread nature of the stacking phenomenon. Significantly, the prochlorophytes, which may serve as models for the evolutionary ancestry of the green algal chloroplast, do not appear to undergo a state transition, indicating that they do not have a mechanism for regulating energy distribution between the photosystems.

This observation, when taken along with our previous arguments with respect to the role of stacking in establishing lateral heterogeneity (Miller & Lyon, 1985), leads us to suggest that thylakoid stacking and the regulation of light-harvesting pigment location are fundamentally different phenomena. It seems reasonable to suggest that the LHC-II phosphorylation/migration system described for higher plants (Staehelin & Arntzen, 1983) may have evolved as a way around the problem of lateral heterogeneity which stacking presents. Therefore, as we suggested earlier (Miller & Lyon, 1985) we have yet to provide a satisfactory answer to the question of why photosynthetic membranes stack.

We find it interesting that Murphy (1986) has suggested that "the possibility of energy transfer between LHC-II units in adjacent appressed membranes ought not to be ignored." While it will be difficult to investigate and explore such interesting possibilities, it is also clear that the algae provide us with a wealth of richly adapted experimental systems in which to pose and answer such questions.

ACKNOWLEDGEMENTS

We are grateful to Mary Kay Lyon, Jules Jacob, Jenny Hinshaw, and Marvyn Steele, whose thoughts and insights have aided in the development of the ideas presented here. This research has been supported by a grant from the National Institutes of Health (GM 28799).

REFERENCES

Allen J F (1983). Protein phosphorylation - Carburettor of photosynthesis? Trends Biochem Sci 8:369-373.

Anderson J, Andersson B (1982). The architecture of photosynthetic membranes: lateral and transverse organization. Trends Biochem Sci 7:288-292.

Biggins J, Bruce D (1985). Mechanism of the light state transition in photosynthesis. III: Kinetics of the state transition in *Porphyridium cruentum*. Biochim Biophys Acta 806:230-236.

Biggins J, Bruce D (1987). The relationship between protein kinase activity and chlorophyll a fluorescence changes in thylakoids from the cyanobacterium *Synechococcus* 6301. in Biggins J (ed): "Progress in Photosynthesis Research," Dordrecht: Martinus Nijhoff, pp 2: 773-776.

Biggins J, Campbell CL, Bruce D (1984) Mechanism of the light state transition in photosynthesis. II: Analysis of phosphorylated polypeptides in the red alga *Porphyridium cruentum*. Biochim Biophys Acta 767:138-144.

Boekema EJ, Dekker JP, van Heel MG, Rögner M, Saenger W, Witt I, Witt HT (1987). Evidence for a trimeric organization of the photosystem I complex from the thermophilic cyanobacterium *Synechococcus* sp. FEBS Letters 217:283-286.

Branton D, Bullivant S, Gilula NB, Karnovsky MJ, Moor H, Mühlethaler K, Northcote DH, Packer L, Satir B, Satir P, Speth V, Staehelin LA, Steere RL, Weinstein RS (1975). Freeze-etching nomenclature. Science 190:54-56.

Burger-Wiersma T, Veenhuis M, Korthals HJ, Van de Wiel CCM, Mur LR (1986). A new prokaryote containing chlorophylls a and b. Nature 320:262-264.

Bullerjahn GS, Matthijs HCP, Mur LC, Sherman LA (1987). Chlorophyll-protein composition of the thylakoid membrane from *Prochlorothrix hollandica*, a prokaryote containing chlorophyll b. Eur J Biochem. 168:295-302.

Deisenhofer J, Epp O, Miki K, Huber R, Michel H (1985). Structure of the protein subunits in the photosynthetic reaction centre of *Rhodopseudomonas viridis* at 3Å resolution. Nature 318:618-624.

Duarte DM, Vesk M (1983). A freeze-fracture study of Cryptomonad thylakoids. Protoplasma 117:130-141.

Gantt EH (1989). *Porphyridium* as a Red Algal Model for Photosynthesis Studies. In Coleman AW, Goff LJ, Stein-Taylor JR (eds): "Algae as Experimental Systems," New York: Alan R. Liss, Inc. pp 249-268.

Gantt EH, Edwards MR, Provasoli L (1971). Chloroplast structure of the Cryptophyceae. Evidence for phycobiliproteins within intrathylakoidal space. J Cell Biol 48:280-290.

Giddings T H, Wasmann C, Staehelin LA (1983). Structure of the thylakoids and envelope membranes of the cyanelles of *Cyanophora paradoxa*. Plant Physiol 71:409-419.

Giddings TH, Withers NW, Staehelin LA (1980). Changes in the thylakoid structure of stacked and unstacked regions of *Prochloron* sp., a prokaryote. Proc Nat Acad Sci U S A 77:352-356

Goodenough UW, Staehelin LA (1971). Structural differentiation of stacked and unstacked chloroplast membranes. J Cell Biol 48:594-613.

Grossman AR, Anderson LK, Conley PB, Lemauc PG (1989). Molecular analyses of Complementary Chromatic Adaptation and Biosynthesis of a Phycobilisome. In Coleman AW, Goff LJ, Stein-Taylor JR (eds): "Algae as Experimental Systems," New York: Alan R. Liss, Inc. pp 269-288.

Ingram K, Hiller RG (1983). Isolation and characterization of a major chlorophyll a/c$_2$ light-harvesting protein from a *Chroomonas* species. Biochim Biophys Acta 722:310-319.

Lewin RA (1975). A marine Synechocystis (Cyanophyta, Chroococcales) epizoic on ascidians. Phycologia 14:153-160.

Miller KR (1979). The structure of a bacterial photosynthetic membrane. Proc Nat Acad Sci U S A 76: 6415-6419.

Miller KR (1981) Freeze-etching studies of the photosynthetic membrane. In Griffith JD (ed): " Electron Microscopy in Biology, volume I." New York: Wiley-Interscience. pp. 1-30.

Miller KR (1982). Three dimensional structure of a bacterial photosynthetic membrane. Nature 300:53-55.

Miller KR (1987) The structure of one photosynthetic membrane. In Miller KR (ed): "*Advances in Cell Biology*, Volume I." New York: JAI Press, pp. 131-156.

Miller KR, Lyon MK (1985). Do we really know why chloroplast membranes stack? Trends Biochem Sci 10:219-222.

Miller KR, Staehelin LA (1976). An analysis of the thylakoid outer surface: coupling factor is limited to unstacked membrane regions. J Cell Biol 68:30-47.

Murphy DJ (1986). The molecular organization of the photosynthetic membranes of higher plants. Biochim Biophys Acta 864:33-94.

Schuster G, Owens GC, Cohen Y, Ohad I (1984). Thylakoid polypeptide composition and light-independent phosphorylation of the chlorophyll *a,b*-protein in *Prochloron*, a prokaryote exhibiting oxygenic photosynthesis. Biochim Biophys Acta 767:596-605.

Snyder UK, Biggins J (1987). Excitation-energy distribution in the cryptomonad alga *Cryptomonas ovata*. Biochim Biophys Acta 892:48-55.

Spear-Bernstein L, Miller KR (1985) Are the photosynthetic membranes of cryptophyte algae inside out? Protoplasma 129:1-9.

Spear-Bernstein, L., Miller, K. (1987) Immunogold localization of the phycobiliprotein of a cryptophyte alga to the intrathylakoidal space. in Biggins J (ed): "Progress in Photosynthesis Research," Dordrecht: Martinus Nijhoff, pp 2: 309-312.

Seibert M, DeWitt M, Staehelin LA (1987). Structural Localization of the O_2-Evolving Apparatus to Multimeric (Tetrameric) Particles on the lumenal Surface of Freeze-etched Photosynthetic Membranes. J Cell Biol 105:2257-2265.

Staehelin LA, Arntzen CJ (1983). Regulation of chloroplast membrane function: Protein phosphorylation changes the spatila organization of photosynthetic membranes. J Cell Biol 97:1327-1337.

Van Niel CB (1935). Photosynthesis of bacteria. Cold Spring Harbor Symp Quant Biol. 3: 138-150.

Wollman F-A, Lemaire C (1988). Studies on kinase-controlled state transitions in photosystem II and b_6f mutants from *Chlamydomonas reinhardtii* which lack quinone-binding protein. Biochim et Biophys Acta 933:85-94.

Algae as Experimental Systems pages 249–268
© **1989 Alan R. Liss, Inc.**

PORPHYRIDIUM AS A RED ALGAL MODEL FOR PHOTOSYNTHESIS
STUDIES

Elisabeth Gantt
Smithsonian/Botany Department, University of
Maryland, College Park, Maryland 20742

Certain microalgae, usually those which can grow well
in defined culture media under controlled conditions, have
emerged as representative species for comparing
morphological and physiological characteristics among
various algal groups. Foremost among those used for
studies on the photosynthetic apparatus are Euglena
gracilis, Chlamydomonas reinhardtii, and Chlorella
pyrenoidosa. Among the red algae, Porphyridium purpureum
(Syn: P. cruentum) is almost the only species which could
be regarded as having a similar status. With the purpose
of fostering further studies on this organism, I will
attempt to summarize most of the current and some past
studies relating to its photosynthetic activity and the
structural features of its photosynthetic apparatus.

Porphyridium's role as a model alga began over thirty
years ago when Duysens (1952), and French and Young (1952)
showed that phycobiliproteins are major light absorbers
for photosynthesis. Later Brody and Vatter (1959)

Abbreviations: APC - allophycocyanin; Chl - chlorophyll;
LCM - high molecular weight terminal pigment; kDa -
kilodalton; PE B and b large and small bangiophyceaen
phycoerythrin; PCB - phycocyanobilin; PEB -
phycoerythrobilin; PUB - phycourobilin; PSI, PSII -
photosystem I, photosystem II, respectively; PIIP -
photosystem II phycobilisome particles; R-PC - rhodophytan
phycocyanin.

examined the cell structure by electron microscopy, and
Brody and Emerson (1959) determined the photosynthetic
pigment composition of cells grown under variable
wavelengths and variable photon flux densities. Murata
(1969) then published his observations illustrating that
in Porphyridium light energy distribution between PSI and
PSII was controlled by light quality (green or red light).

CELLULAR CHARACTERISTICS AND GROWTH

   The unicellular genus Porphyridium is relatively
primitive, and appears to reproduce by vegetative cell
division. The three species are distinct from one another
in pigmentation and their widely different ionic growth
requirements. Strains of P. purpureum have been isolated
from high saline environments, whereas P. sordidum
survives best in diluted seawater, and P. aerugineum grows
in a freshwater medium. Differential phycoerythrin (PE)
concentration accounts for the color variations among the
three species. Phycoerythrin is absent in the blue-green
colored P. aerugineum, but is present in moderate amounts
in the brownish-green colored P. sordidum, and is the
major phycobiliprotein in the brownish-red colored
P. purpureum (cf. Gantt et al. 1979).

   Compared to P. purpureum the other two species grow
relatively slowly, but all grow well in defined media in
batch cultures when continuously shaken and supplied with
1 or 5% CO_2. For P. purpureum the best growth medium is
the artificial sea water medium devised by Jones et al.
(1963). The growth rate of P. purpureum is greatly
influenced by the photon flux density (Table 1). Cell
division is faster in high light than in low light,
although if the light level is raised above ca.
200 μ Einstein m^{-2}s^{-1} growth may be adversely affected.
Low light grown cells have a low compensation point (Table
2), indicating that photosynthesis generates little
additional energy beyond that required for respiration.
Studies by Jahn et al. (1984) also show that there is an
increase of growth at increasing light levels, and that
growth ceases below 220 μWcm^{-2}.

TABLE 1.   Growth of <u>Porphyridium</u> <u>purpureum</u> at three light
photon flux densities*   (Data from Levy & Gantt, 1988)

Culture condition	Growth constant K	Mean gen. time (h) T	Cell density ($\times 10^6$ cell ml^{-1})	Lag period (days)
Low light (10 $\mu E \cdot m^{-2} \cdot s^{-1}$)	0.23	72	2.5	4.0
Medium light (35 $\mu E \cdot m^{-2} \cdot s^{-1}$)	0.46	36	6.5	3.0
High light (180 $\mu E \cdot m^{-2} \cdot s^{-1}$)	0.91	18	15	1.0

*Daylight fluorescence lamps, 18°C.

Growth of P. <u>purpureum</u> in 0.5M glycerol has been
reported in darkness with a significant increase in cell
number (Cheng & Antia, 1970). Interestingly, the
photosynthetic competency does not appear to be diminished
after long dark periods. Growth is resumed within an hour
after illumination following 8 weeks in darkness in a
seawater medium (Bisalputra & Antia, 1980). Although
glycerol serves as a carbon source, it is not sufficient
to support extensive continuous growth. A reduction of
the photosynthetic apparatus to the proplastid level has
not been found; thus limiting developmental studies of the
photosynthetic apparatus. The compensation points and the
light saturation points of P. <u>purpureum</u> are comparable to
those of many other algae (Richardson et al., 1983),
including Antarctic algae (Drew, 1977). Similar
physiologically limiting mechanisms probably occur in most
algae, and reds appear not to be better suited for growth
at lower irradiance levels than any other algal class.

TABLE 2. Photosynthetic characteristics of Porphyridium
purpureum cells grown at varied photon flux densities
(Data from Levy & Gantt, 1988)

| Characteristics | Light Growth | | |
	Low	Medium	High
Sat. intensity $(\mu E \cdot m^{-2} \cdot s^{-1})$	100	138	208
Max. photosynth. $(\mu M\ O_2 \cdot mg\ Chl \cdot h^{-1})$	210	262	350
Dark respiration $(\mu M\ O_2 \cdot mg\ Chl \cdot h^{-1})$	17	23	34
Compensat. point $(\mu E \cdot m^{-2} \cdot s^{-1})$	2	6	19

The unicellular nature of P. purpureum, and the fact
that it lacks a rigid cell wall and fractures easily, are
desirable features for a model organism. A major
undesirable feature is that genetic manipulations are not
possible since there are no known sexual stages. A single
chloroplast is centrally located within the cell (Fig. 1).
The chloroplast thylakoids are often linearly arranged,
and have phycobilisomes attached on the stromal surface.
Starch granules, when present occur in the cytoplasm and
not within the chloroplast. Cells grown in high light
have more dispersed thylakoids and have numerous large
vacuoles in the cytoplasm.

PHOTOSYNTHETIC PIGMENTS AND PHYCOBILISOME STRUCTURE

Phycobilisomes of P. purpureum are hemi-ellipsoidal
in shape with a basal diameter of 40-50 nm and a height
and width of ca. 30-32 nm in measurements made on

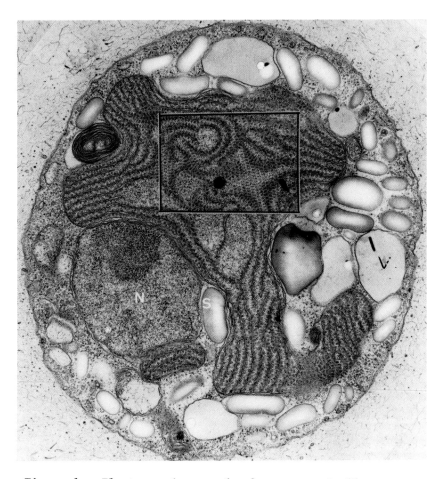

Figure 1. Electron micrograph of an acentrically sectioned cell of P. purpureum (culture grown at ca. 30 $\mu E \cdot m^{-2} \cdot s^{-1}$). The phycobilisomes (in the box) appear in highly ordered rows. Starch (S), nucleus (N), vacuoles (V) are in cell periphery. (25,000 Magnification) (From Gantt & Conti, 1965).

negatively stained preparations (Fig. 2). These phycobilisomes are larger than those generally found in red algae, including P. aerugineum and P. sordidum, and most cyanobacteria which have hemidiscoidally-shaped phycobilisomes that are much smaller (Gantt, 1981; Mörschel & Rhiel, 1987).

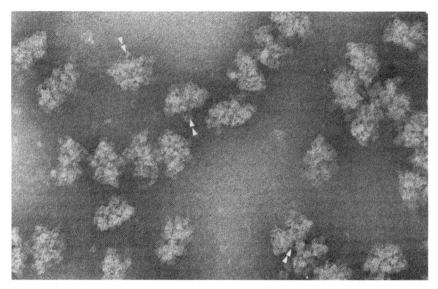

Figure 2. Isolated phycobilisomes negatively stained: associations tend to occur along flattened basal side (arrows). (280,000 Mag.).

Intact phycobilisomes were first isolated from P. purpureum and required glutaraldehyde fixation (cf. Gantt, 1975; 1981). Later a high phosphate medium was devised, which when coupled with detergent solubilization of thylakoids, allows purification of mg quantities of phycobilisomes in a few hours. Phycobilisome isolation is routinely carried out at room temperature by suspending cells in phosphate buffer (ca. 0.7 M, pH 7.0) and breaking them in a French pressure cell, or by sonication. Phycobilisomes are then released by solubilizing the thylakoids with a nonionic detergent. Phycobilisomes remain in the supernate, following removal of large particles by centrifugation. Purified phycobilisomes can be obtained by centrifugation of this supernate on a sucrose-phosphate gradient (0.25-2.0 M sucrose) (Gantt,

1980). Depending on the organism, the molarity of the phosphate and pH may be varied. Phycobilisomes have been isolated from dozens of species by modifications of this procedure. Low temperature (4°-10°C) may lead to uncoupling of energy transfer and partial loss of allophycocyanin. Preparations of phycobilisomes remain stable for days, even weeks, if stored concentrated in phosphate buffer. Contamination from thylakoid remnants can be minimized by a 12-15 h incubation of isolated phycobilisomes in Triton X-100 (20°C).

The phycobilisomes are composed of phycobiliproteins which absorb light of various wavelengths, and linker polypeptides (Glazer, 1985; Mörschel & Rhiel, 1987). The phycobiliproteins have covalently bound chromophores, whereas linker polypeptides are usually colorless. An exception are the gamma subunits of B-PE (Table 3). The linker polypeptides are believed to function by linking the phycobiliproteins in such a way as to assure the proper orientation of the chromophores for maximal energy transfer (Scheer 1986). In vivo most phycobiliproteins exist as oligomers with apparent molecular weights of up to 260 kDa, which when dissociated consist of polypeptides ranging in size from ca. 15-22 kDa. The spectra shown in Fig. 3, show that the phycobiliproteins of P. purpureum greatly extend the absorbance range, and very effectively fill in where Chl absorption is low.

A double absorption maximum at 545 and 563 nm is characteristic of both PE types (B-PE and b-PE), but only B-PE has a significant shoulder at 498 nm (Table 3). Phycoerythrobilin is the major chromophore type in both PEs, while PUB occurs only on the gamma subunit (Glazer, 1977). R-phycocyanin contains both PEB as characterized by absorption at 553 nm, and PCB with absorption at 617 nm. This is relatively unique because most phycocyanins contain only the latter chromophore. Allophycocyanin with a maximum at 650 nm contains only PCB.

Phycoerythrin accounts for over 70% of the phycobilisome protein content, whereas R-PC, APC, and the linker polypeptides account for ca. 9, 5 and 15%, respectively. From the size and pigment composition it is

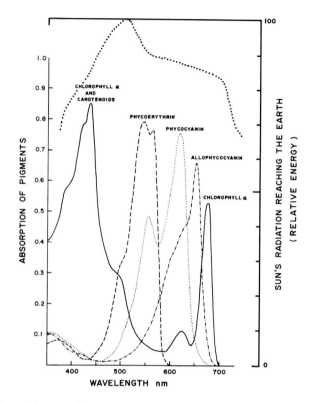

Figure 3. Absorption spectra of major photosynthetically active pigments from P. purpureum and spectrum of solar energy output (upper dotted curve). (From Gantt, 1975).

estimated that a phycobilisome of P. purpureum contains ca. 2000 or more chromophores (Gantt, 1980; Dilworth & Gantt, 1981).

In solution phycobiliproteins are highly fluorescent, and the fluorescence emission property of each phycobiliprotein is very distinctive (Table 3). In the intact phycobilisome little loss of energy occurs from the major phycobiliproteins, but the main fluorescence emission peak comes from terminal pigments. One is a 92 kDa blue polypeptide (L_{CM}), and another by analogy with

cyanobacteria is from a special APC-B polypeptide.
Fluorescence measurements made in steady state condition
show the phycobilisome emission is ca. 675 nm (685 nm at -
196 C) (cf. Gantt et al., 1979; Gantt 1986). By time
resolved fluorescence spectra the energy transfer can be
followed from PE to R-PC to APC to LCM to PS II (Yamazaki
et al., 1984).

TABLE 3. Phycobiliproteins of P. purpureum

Type	Ab Max	Em Max	kDa	Subunits	Chromophores
B-PE	498,545, 563	575	260	$(\alpha\beta)6\gamma$	α2PEB,β3PEB, γ2PEB,2PUB
b-PE	545, 563	575	---	$(\alpha\beta)$n	α2PEB,β3PEB
R-PC	553,615	640	130	$(\alpha\beta)$3	α1PCB, β1PCB, 1PEB
APC	650	660	105	$(\alpha\beta)$3	α1PCB,β1PCB
APC-B	618,673	680	(<210 in PBS core)		(1PCB)
L_{CM}	650	680	92	1	1-2 PCB

Compiled from references by Gantt 1981; Glazer 1977;
Ley et al., 1977; Redlinger & Gantt, 1981.

Polypeptide and spectral analysis of the
phycobilisomes of P. purpureum provide the first evidence
for the L_{CM}'s role as terminal pigment (Redlinger & Gantt,
1981). On SDS-PAGE it retains its blue color, and when
excised from the gel has a fluorescence emission of ca.
680 nm. It is unique in being the largest polypeptide
(Fig. 5) and by its amino acid sequence. Amino acid
sequence analysis of its N-terminal end shows that it is
highly homologous with the 94 kDa polypeptide from Nostoc
sp. but is distinct from other phycobiliproteins (Gantt et

al., 1988). The most definitive evidence for L_{CM} being a main conduit between the bulk phycobiliproteins and PSII has been provided from studies on Nostoc (Mimuro et al., 1986). The role of ACP-B is not clear, but its role as a main conduit is less likely.

The arrangement of the phycobiliproteins within the phycobilisome has been determined from spectral analysis of intact and selectively dissociated phycobilisome, and by immnolabelling at the electron microscope level (cf. Gantt, 1980; 1981; 1986). Phycoerythrin is on the periphery followed by R-PC with APC and the L_{CM} in the core (Fig. 4) and is consistent with the energy sequence path. Although the phycobiliprotein arrangement was first determined in P. purpureum, the complete phycobilisome structure remains to be elucidated. Whereas, several cyanobacterial phycobilisomes have been fully analyzed this is yet to be accomplished for a red alga. This is an interesting challenge, because the composition is complex involving many linker polypeptides (Fig. 5). Especially interesting will be the analysis of the core structure, since it can provide information on the 3-dimensional attachment of the peripheral rods and on the phycobilisome-thylakoid attachment site.

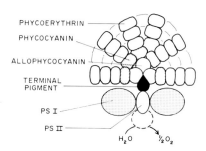

Figure 4. Schematic arrangement of phycobiliprotein organization in P. purpureum phycobilisomes, and presumed association of phycobilisome with the photosystems. The terminal pigment is expected to mediate the excitation energy transfer between the phycobilisome and the PS II antennae chlorophyll complexes.

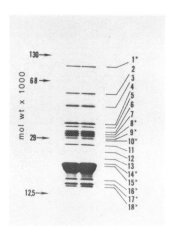
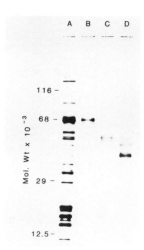

Figure 5 (Left). Polypeptides of P. purpureum
phycobilisomes separated by SDS-PAGE: Bands are numbered
from largest (L_{CM}) to smallest. Chromophore containing
(*) and presumed linkers: Heated (left) unheated (right).
(From Redlinger & Gantt, 1981).
Figure 6 (Right). Polypeptides of major chlorophyll-
protein complexes from P. purpureum. Complexes were first
separated on preparative gels, isolated, denatured, and
re-electrophoresed. Gel stained with Coomassie blue:
(A) unfractioned thylakoids and polypeptides from (B) CPI,
(C) CPIII, (D) CPIV. (From Redlinger & Gantt, 1983).

The role of phycobiliproteins as major light
harvesting pigments is well established from action
spectra, and under physiological conditions most of the
energy absorbed by these pigments is transferred to PSII
(Ley & Butler, 1977). In P. purpureum and several
cyanobacteria it has been determined that up to 95% of the
quanta absorbed by phycobilisomes is first directed to the
Chl-antenna of PSII; however, instead of passing directly
to the reaction center of PSII (RC2), 30 to 50% of the
energy can be delivered to PSI. This transfer is believed
to occur by excitation energy transfer which involves
close contact between PSI and PSII antennae. To gain an
understanding of the interacting mechanism requires
analysis of the composition of the chlorophyll complexes,
and knowledge of their stoichiometry and their topology in
the thylakoid membrane.

From our previous results we know that three Chl-protein complexes (CPI, CPIII, CPIV) can be electrophoretically separated, which on full denaturation reaches into three major molecular weight regions (Fig. 6). CPI contained P700 and had a fluorescence emission at 720 nm (-196 C). In denaturing gels the apoprotein apparent molecular weight was 68 kDa. CPIII and CPIV appear to belong with PSII as suggested by the lack of P700, and from their fluorescence maximum at 690 nm. On complete denaturation CPIII had a predominant band at ca. 52 kD, which according to other systems is probably a PSII core antenna polypeptide. Sometimes a 92 kD component is present as a minor band in CPIII. CPIV when completely denatured contained some of the 52 kD, but the major constituents were polypeptides of 40 and 48 kD (Fig. 6).

CPI contained ca 41% of total Chl, with CPIII and CPIV having 3 and 4% respectively, and the remainder being free Chl (presumably released by detergent action required for separation of the complexes). CPI represents a major portion of the Coomassie stainable thylakoid protein and contains most of the Chl. This is also consistent with the results from cyanobacteria where the chlorophyll content of PSI is greater than in PSII (cf. Anderson & Barrett, 1986; Gantt, 1986).

Whereas a spatially close relationship between PSII and PBS is expected from action spectra and fluorescence determinations, this association was only recently established by the isolation of PIIP particles (Chereskin et al., 1985; Clement-Metral et al., 1985). These particles are phycobilisomes plus a small amount of Chl and retention of the O_2-evolving complex. They have an average of 90 Chl/PBS compared to 1200 Chl/PBS in unfractionated thylakoids. P700 was not detected in these particles, and they lacked the PSI apoproteins. Furthermore, functional coupling between the PBS and PSII was established, since green light (absorbed by PE, 84% of P. cruentum PBS phycobiliprotein) was more effective in promoting PSII activity than wavelengths shorter than 665 nm.

ACCLIMATION TO LIGHT

Of particular interest are the changes occurring within the photosynthetic apparatus during acclimation to light. Laboratory studies on several red algal species have shown that at lower irradiances (around 10 $E \cdot m^{-2} \cdot s^{-1}$) the cell pigment content is greater than at higher irradiances (around 150-200 $E \cdot m^{-2} \cdot s^{-1}$) (Levy & Gantt, 1988; Waaland et al., 1974). Similar trends are reflected in field collected species (Larkum & Barrett, 1983; Rosenberg & Ramus, 1982). At what point does the acclimation occur? Is there a change in the ratio of PSI:PSII phycobilisomes? A change in the antennae size of PSII was reported by Waaland et al. (1974) in the red alga Griffithsia pacifica. They observed a lower PE content, relative to Chl, at higher light intensity, along with a reduction in phycobilisomes per thylakoid area. Similar results were obtained for P. purpureum (Levy & Gantt, 1988) in batch cultures with cell densities up to 15 x 10^6 cell ml^{-1} in high light.

TABLE 4. Pigment content of P. purpureum cells grown at three light photon flux densities. (Data from Levy & Gantt, 1988).

Culture condition	PE (x10^{-6} $\mu g \cdot cell^{-1}$) ± SD	Chl	Molar ratio Chl/P700 ± SD
Low light (10 $\mu E \cdot m^{-2} \cdot s^{-1}$)	23±5	2.2±0.6	111±10
Medium light (35 $\mu E \cdot m^{-2} \cdot s^{-1}$)	20±7	2.0±0.4	110±8
High light (180 $\mu E \cdot m^{-2} \cdot s^{-1}$)	10±3	1.8±0.4	95±8

As shown, cells grown at the highest irradiance had the highest photosynthetic capacity (Tables 1, 2) and the

greatest growth rate. However, as seen in Table 4, the
Chl content per cell remained relatively unchanged as did
Chl/P700, but the PE content and phycobilisome number (not
shown) were reduced by 50%. Whereas, the photosynthetic
unit size, i.e. the total pigment content (chlorophyll
plus phycobiliproteins) was clearly reduced, what accounts
for increased photosynthetic capacity? Jahn et al. (1984)
had already shown that ribulose bisphosphate carboxylase
activity is not affected by irradiance. It is possible
that the PSII reaction center number per cell, which is
not yet known, may be greater in high irradiance cells.
It is also possible that the PSII reaction centers in
these cells have different turnover rates.

Staehelin et al. (1978) in an electron microscopy
study, including freeze fracturing, of Griffithsia
pacifica also found a decrease in phycobilisome number at
high irradiance cells, but in addition attempted to obtain
some measure of the PSII number per thylakoid area. For
freeze-fractured thylakoids they counted the 10 nm
particles on the exoplasmic surface, which we now know
represent the PSII core particle (Mörschel & Schatz,
1987). They noted only small differences in high and low
light cells. However, their results suggested that about
three to four PSII particles could be associated with each
phycobilisome. From photophysical measurements, Ley
(1984) also concluded that three to four PSII reaction
centers can be functionally linked to each phycobilisome.
Similarly, a ratio of four PSII reaction centers per
phycobilisome have been found in another red alga
Neoaghardhiella bailyei (Kursar & Alberte, 1983). What
functional advantage there would be for an association of
multiple PSII reactions per phycobilisome remains to be
elucidated. Multiple connections with PSII may assure
maximal excitation energy transfer from phycobilisomes to
the Chl antennae of PSII, and may even enhance the
probability of "spilloner" to PSI.

Finally, the above results lead to the consideration
of the possible structural arrangement of the
phycobilisomes with PSI and PSII in the thylakoid. A
diagramatic presentation in Fig. 7 of the possible
arrangement of the photosystem around phycobilisomes,
although this model is yet to be tested, it is supported
by structural and physiological data. The model is

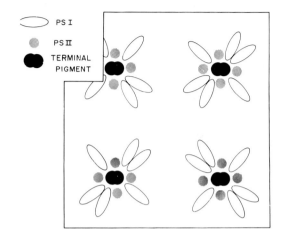

PS I
PS II
TERMINAL
PIGMENT

Figure 7. Possible arrangement of the PSII mean base of phycobilisome and PSI peripheral to PSII. Such an arrangement allows for possible rapid changes required in excitation energy transfer, and assumes that the phycobilisomes and chlorophyll antennae complexes of PSI and PSII are functionally linked and have a homogenous topography.

significantly different from those proposed for higher plants in that PSI and PSII are spatially closely associated and are not sequestered into separate regions. The mechanism for regulation of excitation energy distribution appears to be different from grana-containing plants. According to Biggins et al. (1984) reversible protein phosphorylation does not appear to be involved in Porphyridium. If the photosystems are closely associated around the phycobilisomes, then small conformation changes at the base of the phycobilisome can cause a change in the distribution of excitation energy. The basis for such a model is detailed in Biggins et al. (1984).

Interestingly, changes in color from pink to red were observed in whole plants of the red alga Callithamnion by Yu et al. (1981). On extracting the phycobiliproteins they found that the color changes were due to differences

in the PE composition, rather than a change in the ratio
of PE:PC:APC. The PE had variable ratios of PUB:PEB.
They determined that this was attributable to changes in
light intensity and not light quality, and thus is not an
example of complementary chromatic adaptation as occurs in
some cyanobacteria. Although light quality has been shown
to affect the phycobiliprotein to chlorophyll ratio in
P. purpureum (Ley & Butler, 1980), in whole cell spectra,
there was no detectable variation in the ratio of the
phycobiliprotein types. Thus, whereas certain pigment
changes occur, there is no clear evidence for
complementary chromatic adaptation in red algae.

CONCLUSION

     As a model organism for studies on the photosynthetic
apparatus of red alga, P. purpureum has many favorable
features. It is a unicellular alga and can be easily
grown under defined conditions. Its physiological
responses to variable photon flux densities have been
described. The major features of the phycobilisome
structure and the Chl-binding complexes have been
isolated. With this model system it is now possible to
address significant problems on how the photosynthetic
apparatus is regulated by light quality and intensity.
Major questions which are being addressed relate to the
stoichiometry between PSI, PSII, and phycobilisomes; the
topography of the photosystems and energy distribution;
and the mechanism by which light regulation occurs.

Acknowledgement is made of support by U.S. Dept. of Energy
Contract AS05-76ER-04310 and Grant DE-FG05-87ER13652.

REFERENCES

Anderson JM, Barrett J (1986). Light-harvesting pigment-
     protein complexes of algae. In Staehelin LA, Arntzen CJ
     (eds): "Encyclopedia of Plant Physiology. Vol 19.
     Photosynthetic Membranes and Light-Harvesting Systems,"
     Berlin: Springer-Verlag, pp 269-285.
Biggins J, Campbel CL, Bruce D (1984). Mechanism of the
     light state transition in photosynthesis. V. Analysis of

phosphorylated polypeptides in the red alga Porphyridium cruentum. Biochim Biophys Acta 767:138-144.

Bisalputra T, Antia NJ (1980). Cytological mechanisms underlying darkness-survival of the unicellular red alga Porphyridium cruentum. Bot Marina 23:719-730.

Brody M, Emerson R (1959). The effect of wavelength and intensity of light on the proportion of pigments in Porphyridium cruentum. Am J Bot 46:433-440.

Brody M, Vatter AE (1959). Observations on cellular structure of Porphyridium cruentum. Biophysic Biochem Cytol 25:289-293.

Cheng, JY, Antia NJ (1970). Enhancement by glycerol of photoautotrophic growth of marine planktonic algae and its significance to ecology of glycerol pollution. J Fish Res Bd Canada 27:335-354.

Chereskin B, Clement-Metral JD, Gantt E (1985). Characterization of a purified photosystem II-phycobilisome particle preparation from Porphyridium cruentum. Plant Physiol 77:626-629.

Clement-Metral, JD, Gantt E, Redlinger T (1985). A photosystem II-phycobilisome preparation from the red alga Porphyridium cruentum: oxygen evolution, ultrastructure, and polypeptide resolution. Arch Biochem Biophys 238:10-17.

Dilworth MF, Gantt E (1981). Phycobilisome-thylakoid topography on photosynthetically active vesicles of Porphyridium cruentum. Plant Physiol 67: 608-612.

Drew EA (1977). The physiology of photosynthesis and respiration in some antarctic marine algae. Br Antarct Surv Bull 46:59-76.

Duysens LNM (1952). Transfer of Excitation Energy in Photosynthesis. Ph.D. Thesis, University of Utrecht, Holland pp. 104.

French CS, Young CK (1952). The fluorescence spectra of red algae and their transfer of energy from phycoerythrin to chlorophyll. J Gen Physiol 35:873-890.

Gantt E (1975). Phycobilisomes: light harvesting pigment complexes. BioScience 25:781-788.

Gantt E (1980). Structure and function of phycobilisomes: light harvesting pigment complexes in red and blue-green algae. Int Rev Cytol 66:45-80.

Gantt E (1981). Phycobilisomes. Ann Rev Plant Physiol 32:327-347.

Gantt E (1986). Phycobilisomes. In Staehelin LA, Arntzen CJ (eds): "Encyclopedia of Plant Physiology. Vol 19.

Photosynthetic Membranes and Light Harvesting Systems,"
Berlin: Springer-Verlag, pp 260-268.
Gantt E, Conti SF (1965). The ultrastructure of
Porphyridium cruentum. J Cell Biol 26:365-381.
Gantt E, Cunningham FX, Lipschultz CA, Mimuro M (1988).
N-terminus conservation in the terminal pigment of
phycobilisomes from a prokaryotic and eukaryotic alga.
Plant Physiol 86:996-998.
Gantt E, Lipschultz CA, Grabowski J, Zimmerman BK (1979).
Phycobilisomes from blue-green and red algae. Isolation
criteria and dissociation characteristics. Plant
Physiol 63:615-620.
Gantt E, Lipschultz CA, Redlinger T (1985).
Phycobilisomes: a terminal acceptor pigment in
cyanobacteria and red algae. In Steinback K, Bonitz S,
Arntzen CJ, Bogorad L (eds): "Molecular Biology of the
Photosynthetic Apparatus," Cold Spring Harbor: Cold
Spring Harbor Laboratory, pp 223-229.
Glazer AN (1977). Structure and molecular organization of
the photosynthetic accessory pigments of cyanobacteria
and red algae. Mol Cell Biochem 18:125-140.
Glazer AN (1985). Light harvesting by phycobilisomes.
Annu Rev Biophys Biophys 14:47-77.
Jahn W, Steinbiss J, Zetsche K (1984). Light intensity
adaptation of the phycobiliprotein content of the red
alga Porphyridium. Planta 161:536-539.
Jones RH, Speer HL, Kurry W (1963). Studies on the growth
of the red alga Porphyridium cruentum. Physiol Plant
16:636-643.
Kursar T, Alberte R (1983). Photosynthetic unit
organization in a red alga. Relationships between
light-harvesting pigments and reaction centers. Plant
Physiol 72:409-414.
Larkum AWD, Barrett J (1983). Light-harvesting processes
in algae. In Woolhouse HW (ed): "Advances in Botanical
Research. Vol 10," New York: Academic, pp 3-219.
Levy I, Gantt E (1988). Light acclimation of Porphyridium
purpureum (Rhodophyta): growth, photosynthesis, and
phycobilisomes. J Phycol 24:452-458.
Ley AC (1984). Effective absorption cross-sections in
Porphyridium cruentum. Implications for energy transfer
between phycobilisomes and photosystem II reaction
centers. Plant Physiol 74:451-454.
Ley AC, Butler WL (1977). The distribution of excitation
energy transfer between photosystem I and photosystem II
in Porphyridium cruentum. In Miyachi S, Katoh K, Fujita

Y, Shibata K (eds): "Photosynthetic Organelles. Special Edition Plant Cell Physiol," Tokyo: Japanese Society of Plant Physiologists, pp 33-46.

Ley AC, Butler WL (1980). Effects of chromatic adaptation on the photochemical apparatus of photosynthesis in Porphyridium cruentum. Plant Physiol 65:714-722.

Ley AC, Butler WL, Bryant DA, Glazer AN (1977). Isolation and function of allophycocyanin B of Porphyridium cruentum. Plant Physiol 59:974-980.

Mimuro M, Lipschultz CA, Gantt E (1986). Energy flow in the phycobilisome core of Nostoc sp. (Mac): two independent terminal pigments. Biochim Biophys Acta 852:126-132.

Mörschel E, Schatz (1987). Correlation of photosystem II complexes with exoplasmatic freeze-fracture particles of thylakoid of the cyanobacterium Synechococcus. Planta 172:145-154.

Mörschel E, Rhiel E (1987). Phycobilisomes and thylakoids: the light harvesting system of cyanobacteria and red algae. In Harris JR, Horne RW (eds): "Electron Microscopy of Proteins. Vol 6. Membrane Structures," London: Academic, pp 209-254.

Murata N (1969). Control of excitation energy transfer in photosynthesis. I. Light induced change of chlorophyll fluorescence in Porphyridium cruentum. Biochim Biophys Acta 172:242-251.

Redlinger T, Gantt E (1981). Phycobilisome structure of Porphyridium cruentum: polypeptide composition. Plant Physiol 68: 1375-1379.

Redlinger T, Gantt E (1983). Photosynthetic membranes of Porphyridium cruentum: an analysis of chlorophyll-protein complexes and heme-binding proteins. Plant Physiol 73:36-40.

Richardson K, Beardall J, Raven JA (1983). Adaptation of unicellular algae to irradiance: an analysis of strategies. New Phytol 93:157-191.

Rosenberg G, Ramus J (1982). Ecological growth strategies in the seaweeds Gracilaria foliifera (Rhodophyceae) and Ulva sp. (Chlorophyceae): photosynthesis and antenna composition. Mar Ecol Prog Ser 8:233-241.

Scheer H (1986). Excitation transfer in phycobiliproteins. In Staehelin LA, Arntzen CJ (eds): "Encyclopedia of Plant Physiology. Vol 19. Photosynthetic Membranes and Light Harvesting Systems," Berlin: Springer-Verlag, pp 325-337.

Staehelin LA, Giddings TH, Badami P, Krsymowski WW (1978).
A comparison of the supramolecular architecture of
photosynthetic membranes of blue-green, red and green
algae and of higher plants. In Deamer DW (ed): "Light
Transducing Membranes," New York: Academic, pp 335-355.
Waaland JR, Waaland SD, Bates G (1974). Chloroplast
structure and pigment composition in the red alga
Griffithsia pacifica. J Phycol 10:193-199.
Yamazaki I, Mimuro M, Murao T, Yamazaki T, Yoshihara K,
Fujita Y (1984). Excitation energy transfer in the
light-harvesting antennae system of the red alga
Porphyridium cruentum and the blue-green alga Anacystis
nidulans. Analysis of time resolved fluorescence
spectra. Photochem Photobiol 39:233-240.
Yu MH, Glazer AN, Spencer KG, West J (1981).
Phycoerythrins of the red alga Callithamnion, variation
in phycoerythrobilin and phycourobilin content. Plant
Physiol 68:482-488.

Algae as Experimental Systems pages 269–288
© **1989 Alan R. Liss, Inc.**

MOLECULAR ANALYSES OF COMPLEMENTARY CHROMATIC ADAPTATION AND
THE BIOSYNTHESIS OF A PHYCOBILISOME

Arthur R. Grossman, Lamont K. Anderson, P.B. Conley and
P.G. Lemaux

Carnegie Institution of Washington, Department of Plant
Biology, 290 Panama Street, Stanford, California 94305
(A.R.G, L.K.A.), V.A. Medical Center M-151, 3801
Miranda, Palo Alto, California 94304 (P.B.C.) and
Pfizer Inc., Central Research, Groton, Connecticut
06340 (P.G.L.)

Phycobilisomes are light-harvesting complexes associated with photosynthetic membranes of cyanobacteria and
thylakoid membranes in chloroplasts of red algae. They
absorb light energy in the 500-650 nm range and transfer
this energy to membrane-bound chlorophyll protein complexes
(Glazer, 1985). Our current understanding of phycobilisome
structure and function is based on almost 25 years of research since their initial characterization in the red alga
Porphyridium cruentum (Gantt & Conti, 1966a, b). There is
now a comprehensive model for phycobilisome structure and
function (Glazer et al., 1983), detailed characterizations
of the individual polypeptides of the complex (Glazer, 1985)
and considerable information concerning the organization of
genes encoding phycobilisome components and their regulated
expression (Grossman et al., 1986, 1988).
   In many cyanobacteria, phycobilisome structure and
composition changes in response to environmental cues. Nutrient stress triggers the degradation of phycobilisomes,
converting them into metabolites for cell maintenance (Allen

Abbreviations: AP - allophycocyanin; an α or β given as a
superscript to AP, PC or PE indicate specifically the α or β
subunit of that biliprotein; D - dark grown; G - green light
grown; kb - kilobase pair; kDa - kilodalton; $L_C^{7.9}$, $L_R^{9.7}$,
$L_R^{37.5}$, L_R^{39}, are the linker (L) polypeptides of the phycobilisome either located in the core (subscript C) or the
rods (subscript R); PC - phycocyanin; PE - phycoerythrin; R
- red light grown. CIW-DPB Publication No. 1020.

& Smith, 1969). Light quantity also affects phycobilisome levels; high light intensity causing a decrease in the number and size of phycobilisomes whereas low light has an opposite effect (Yamanaka & Glazer, 1981; Lönneborg et al., 1985). The response of certain cyanobacteria to light quality involves altering levels of specific phycobilisome components, thereby optimizing absorption of prevalent wavelengths of available light (Tandeau de Marsac, 1977; Bryant & Cohen-Bazire, 1981). This acclimation process has been termed chromatic adaptation (Bogorad, 1975; Tandeau de Marsac, 1983).

Phycobilisome biosynthesis is the processes beginning with the transcription of genes encoding phycobilisome polypeptides and leading to production of a functional light harvesting complex. Interesting aspects of phycobilisome biosynthesis and degradation include identification and characterization of sensors involved in the perception of light and nutrient levels, mechanisms of transcriptional control and processes involved in the maturation (e.g. chromophore attachment, glycosylation) and assembly of individual phycobilisome polypeptides. Events required for the controlled production of phycobilisome polypeptides and their assembly may be analogous to those involved in the production of other multi-protein structures and therefore the study of phycobilisome biosynthesis will be of global importance. As a model for the construction of a multi-protein complex, phycobilisomes have several attractive properties. They are not essential for cell growth and therefore mutants in structure, assembly, and regulation are readily obtained. They are easy to isolate and extremely abundant, sometimes representing up to 50% of the cellular dry weight. The intact phycobilisome can be dissociated to discrete substructures by changing buffer conditions (Gantt et al., 1976), and reconstitution of the individual components can be achieved. Electron microscopy can be used to directly observe the substructures of the complex. Absorption characteristics of the covalently bound chromophores on the phycobiliproteins are sensitive to the local protein environment and assembly state and therefore serve as powerful indicators of structural and functional features of isolated complexes. Many of the genes encoding phycobilisome structural proteins have been cloned and sequenced, providing the foundation for analyzing regulated gene expression and functional aspects of phycobiliprotein polypeptides. Research on cyanobacterial chromatic adaptation may be particularly relevant since this response shares some features with phytochrome mediated responses of higher plants and will ultimately expand our understanding of photoperceptors in photosynthetic organisms.

PHYCOBILISOME STRUCTURE

Phycobilisomes contain the chromophoric biliproteins and non-chromophoric linker polypeptides organized into two structural domains, the core and the rods (Fig. 1). Three major classes of biliproteins are present in many cyanobacteria; phycoerythrin (PE, λ_{max}= 565nm), phycocyanin (PC, λ_{max}= 617nm) and allophycocyanin (AP, λ_{max}= 650nm). All consist of two dissimilar subunits, α and β, that form monomers. Three monomers form a trimer, $(\alpha\beta)_3$, and in PC and PE, two trimers interact to form hexameric disks. Each disk in the rod (PE or PC) is associated with a single and specific linker polypeptide (Lundell et al., 1981). These polypeptides determine the assembly sequence in the rod substructure and also modulate the spectral properties of the bilin chromophores, facilitating energy transfer to the core. The core is in contact with chlorophyll of the photosynthetic apparatus and contains AP, whereas the rods radiate from the core and contain PC and PE. Recent models of the core substructure have been published (Anderson & Eiserling, 1986; Lundell & Glazer, 1983a,b,c). Within the rods, PC hexamers are proximal to the core whereas PE hexamers are distal and the energy transfer pathway is PE to PC to AP to chlorophyll. The rod structure is more amenable than the core to detailed analyses and the interactions among many of the rod components is elucidated by in vitro reconstitution from individual polypeptides or substructures and x-ray crystallography of PC hexamers (Schirmer et al., 1986). Interactions among biliprotein and linker polypeptides provide a network of bilin chromophores in the complex that is optimized for fast (about 54 picoseconds) and efficient (better than 95%) energy transfer to chlorophyll (Glazer et al., 1985).

Our laboratory is currently investigating phycobilisome biosynthesis in the filamentous and unicellular cyanobacteria Fremyella diplosiphon and Synechocystis 6701. This article will summarize our recent work on differential gene expression during chromatic adaptation, the photobiology of this response and the analysis of mutants which exhibit

F. diplosiphon Phycobilisome Structure

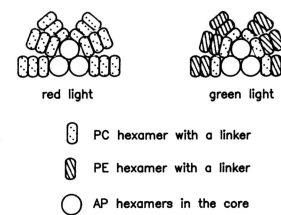

red light   green light

    PC hexamer with a linker

    PE hexamer with a linker

    AP hexamers in the core

Fig. 1. Schematic representation of *Fremyella diplosiphon* phycobilisome structure in R and G.

either aberrant photoregulation or are defective in phycobilisome assembly.

LIGHT REGULATION OF PHYCOBILIPROTEIN EXPRESSION

   A first step toward understanding coordinate and differential regulation of phycobilisome components at the molecular level is the isolation of genes encoding these components. Based on previous results indicating that phycobiliproteins in eukaryotic algae were likely to be plastid encoded (Egelhoff & Grossman, 1983) and that the plastid genome is small relative to the cyanobacterial genome (~150 kb compared to ~5,000 kb) (Palmer, 1985) we decided to isolate phycobiliprotein genes from the plastid genome of the eukaryotic alga *Cyanophora paradoxa*. Using an immunological screening procedure we isolated clones encoding PC^{α} and PC^{β} (Lemaux & Grossman, 1984, 1985). Homologies between the PC and AP genes enabled us to use the PC sequences to isolate the plastid encoded AP gene set (Lemaux & Grossman, 1985). Phycobiliprotein genes have now been isolated from a number of organisms including the eukaryotic alga *Cyanophora para-*

doxa (Bryant <u>et</u> <u>al</u>., 1985b; Lemaux & Grossman, 1984, 1985) and prokaryotic cyanobacteria, *Synechococcus* sp. PCC 6301 (Houmard <u>et</u> <u>al</u>., 1986; Lind <u>et</u> <u>al</u>., 1985), *Synechococcus* sp. PCC 7002 (Bryant <u>et</u> <u>al</u>., 1985a; de Lorimier <u>et</u> <u>al</u>., 1984; Pilot & Fox, 1984), *Synechocystis* sp. PCC 6701 (Anderson & Grossman, 1987) *Anabaena* sp. PCC 7120 (Belknap & Haselkorn, 1987), *Pseudanabaena* sp. PCC 7409 (Dubbs & Bryant, 1987) and *Fremyella diplosiphon* (Conley <u>et</u> <u>al</u>., 1985, 1986, 1988; Mazel <u>et</u> <u>al</u>. 1986, 1988).

Since complementary chromatic adaptation does not occur in eukaryotic algae such as *C. paradoxa*, we focused most of our efforts on *F. diplosiphon*, a chromatically adapting organism. In this cyanobacterium, dramatic changes in the levels of both pigmented and nonpigmented proteins of the phycobilisome occur during light quality acclimation. PE is present at high levels in phycobilisomes from cells grown in G but barely detectable in R-grown cells whereas specific PC subunits (inducible PC) are abundant in R- but absent in G-grown cells. A second set of PC subunits (constitutive PC), present at similar levels in phycobilisomes from both R- and G-grown organisms, serves as an obligatory energy bridge between PE and AP when cells are grown in G. Furthermore, linkers associated with PE are barely detectable in R-grown cells and linkers associated with the inducible PC are undetectable in G-grown cells. The levels of AP and the AP-associated core linker are similar in cells grown in either R or G.

Biliprotein genes from *C. paradoxa* were used for isolating analogous genes from *F. diplosiphon*. Characterization of *F. diplosiphon* clones indicated that genes encoding several phycobilisome components were clustered on the cyanobacterial genome (Conley <u>et</u> <u>al</u>., 1986; Grossman <u>et</u> <u>al</u>., 1988). A 13 kb region of DNA (Fig. 2) contains two sets of PC genes, one expressed only in R (*cpc*B2A2, inducible PC) and the other expressed in both R and G (*cpc*B1A1, constitutive PC), a single AP gene set (*apc*A1B1) and at least four linker polypeptide genes (*apc*C, *cpc*D, *cpc*H and *cpc*I). The two PC gene sets are contiguous and transcribed in the same direction whereas the AP gene set is upstream from *cpc*B2A2, with genes encoding linker polypeptides both 5' and 3' to the AP coding region. The gene *apc*C, which encodes the small linker polypeptide $L_C^{7.9}$ present in the core of the phycobilisome, is immediately downstream of *apc*B$_1$. An open reading frame located upstream of the AP gene set also has homology to linker polypeptides and probably encodes the anchor protein. Genes for the three linker polypeptides, $L_R^{37.5}$, L_R^{39} and $L_R^{9.7}$ (gene designations cpcI, cpcH and cpcD, respectively), which are associated with the inducible PC subunits, are located downstream from the R-inducible cpcB2A2 gene set. An open reading frame (ORF) thought to be involved in phycobilisome biosynthesis is immediately downstream of the constitutive PC gene set and is conserved in sequence and genomic position among several cyanobacterial species (Belknap & Haselkorn, 1987; Conley et al., 1988).

Homologies to phycobiliprotein and linker polypeptide genes were found on other clones as well. One such clone contains the gene set for PE (cpeBA) (Mazel et al., 1987; Grossman et al., 1988). Based on low stringency hybridizations with the linker genes contiguous to cpcB2A2, there are no genes encoding linker proteins adjacent to cpeBA. Another clone that has been isolated contains a third set of PC genes, cpcB3A3 (Mazel et al., 1988; Grossman et al., 1988), and sequences homologous to linker polypeptides (Grossman et al., 1988, see below).

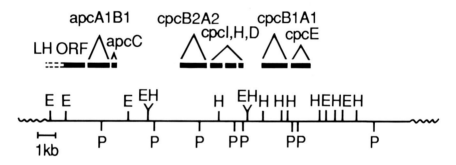

Fig. 2. Restriction endonuclease map of F. diplosiphon chromosomal DNA containing a cluster of genes encoding phycobilisome components. Abbreviations: apcA1B1, gene set for AP α & β; cpcB1A1, gene set for constitutive PC β & α; cpcB2A2, gene set for "inducible" PC β & α; apcC, gene for $L_C^{7.9}$; cpcI, H, D are genes for $L_R^{37.5}$, L_R^{39}, and $L_R^{9.7}$, respectively; cpcE, a 30 kDa ORF; LH ORF, an open reading frame with linker homology; H, HindIII; P, PstI; E, EcoRI. Heavy lines delineate coding regions of the individual genes (Grossman et al., 1988).

Amino acid sequences of phycobiliprotein subunits and linker polypeptides, as derived from the gene sequences, have helped in identifying residues which may be important for protein-protein and protein-pigment interactions. Certain invariant residues of the biliprotein may be required for the attachment and orientation of the chromophores (Schirmer et al., 1986). Conserved regions among the linker polypeptides may be required for glycosylation, protein conformation, and interaction with the biliprotein hexamers.

Using gene specific probes, transcript accumulation from the different phycobiliprotein genes has been examined and the effects of light quality on transcription established (Fig. 3).

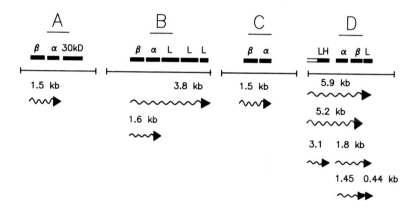

Fig. 3. Transcript analyses of *F. diplosiphon* phycobiliproteins and linker polypeptide genes. A, constitutive PC (*cpc*A1B1); B, "inducible" PC (*cpc*B2A2); C, PE (*cpe*BA); D, AP (*apc*A1B1): wavy arrows define boundaries of transcripts; size of each mRNA is designated above arrow.

The cpcB1A1 gene set is transcribed as a single RNA species of 1,500 bases (panel A) (Conley et al., 1986) present in RNA from both R- and G-grown cells whereas the cpcB3A3 gene set is transcribed, in both R- and G-grown *F. diplosiphon* as a 2,000 base species (Mazel et al., 1988). The abundance of the latter species is ca. 50-fold less than the former. The role of the gene products from cpcB3A3 in the phycobilisome is not known.

The R-inducible cpcB2A2 gene set is transcribed as two different-sized mRNAs (Fig. 3, panel B) detectable in R-grown cells but not in G-grown cells. The smaller transcript (1,600 bases) encodes the two PC subunits whereas the larger (3,800 bases) encodes the two PC subunits plus the three PC linker polypeptides, $L_R^{37.5}$, L_R^{39}, and $L_R^{9.7}$ (Lomax et al., 1987). Thus, the larger transcript probably encodes all of the structural elements required for altering the rod substructure upon shifting F. diplosiphon from G to R. The 3,800 base transcript is present at ca. 15% the level of the smaller one (Conley et al., 1988) which reflects the relative levels of the inducible PC subunits and the linker polypeptides in the phycobilisome from R-grown cells. The 1,600 base transcript may be generated either by nucleolytic processing of the 3,800 base species (segmental stability of the RNA) or transcription termination 85% of the time at the 3-prime end of the 1,600 base transcript (and read through 15% of the time) (Conley et al., 1988).

The cpeBA gene set (Fig. 3, panel C) is transcribed as a 1,500 base mRNA which is abundant in G-grown cells but barely detectable in cells maintained in R. No larger transcript hybridizes to this gene set (unlike with cpcB2A2) and no linker homologies are contiguous.

The apcA1B1 gene set shows a complex transcriptional pattern (Fig. 3, panel D). The α and β subunits of AP are encoded on a 1,450 base RNA species whereas the core linker is encoded on a 440 base species. Sequences contained on the 1,450 and 440 base species are also contained on larger transcripts (1,800 and 5,900 bases). A 5,200-base species is believed to encode the AP subunits but not the core linker polypeptide. Upstream of apcA1B1 is an ORF which has strong homology to the linkers associated with the inducible PC subunits and probably encodes the anchor protein. It is encoded by transcripts of 3,100, 5,200 and 5,900 bases, the latter two also encoding the AP subunits. In all cases, these transcripts are present in equal amounts in cells grown in R and G, an observation consistent with the finding that the phycobilisome core is not altered during acclimation of F. diplosiphon to light quality. Determining the ways in which the different transcripts encoding the core components are generated may provide insights into transcriptional termination and RNA processing in cyanobacteria.

In order to help establish which nucleotide sequences are important for transcriptional regulation during chromatic adaptation, transcription initiation sites for the apcA1B1, cpeBA, cpcB1A1 and cpcB2A2 gene sets have been determined by S1 nuclease digestion (Conley et al., 1988, Grossman et al., 1988; Mazel et al., 1988). Essentially no homology was found between the upstream regions of the two constitutively expressed genes, cpcB1A1 and apcA1B1. Among species, however, there is limited homology in the sequences just upstream of transcription initiation of the PC gene sets (Belknap & Haselkorn, 1987; Pilot & Fox, 1984). Such findings are of limited utility without genetic and biochemical means (e.g., promoter deletions analyses, in vitro transcription, footprinting) to probe promoter function.

Although our data indicates a central role for transcriptional differences in the synthesis of phycobiliproteins and linkers during growth of F. diplosiphon in different light qualities, there may be other regulatory events at the post-transcriptional (RNA processing), translational (codon usage) and post-translational (chromophore attachment, glycosylation) levels. Such processes may serve to fine tune the synthesis of various components in this macromolecular complex.

PHOTOBIOLOGY

Initial studies in our laboratory on the molecular basis of complementary chromatic adaptation have now been extended to encompass the photobiology of this system (collaboration with laboratory of Dr. Winslow Briggs). Cultures of F. diplosiphon were acclimated to a constant fluence rate of R and G and the levels of mRNAs encoding the different phycobilisome components were measured after transfer of R-acclimated cells to G and G-acclimated cells to R. Levels of transcripts for AP subunits, the core linker polypeptide and the constitutive PC subunits are similar in R and G. In contrast, levels of PE and inducible PC mRNAs increased rapidly after transferring F. diplosiphon from R to G and G to R, respectively (Oelmüller et al., 1988a). After growth under our R conditions, the cells contain 5-10% of the amount of PE subunits and PE mRNA observed in G-grown cells. PE mRNA accumulates to maximal levels 4 h after transfer of R-acclimated cells to G, whereas the inducible PC mRNA reaches a maximum 2 h after transfer of G-acclimated cells to R. The levels of the inducible PC transcripts decline rapidly under non-inductive illumination, whereas the decline in PE mRNA levels is much slower [only a 20-30% decrease is observed 8 h following transfer of G-grown cells to R, whereas 2 h after transfer of R-grown cells to G, the inducible PC mRNA is barely detectable, (Oelmüller et al.,

1988a)].

To determine the contribution of synthesis and degradation to absolute levels of phycobiliprotein transcripts, we used rifampicin to block *de novo* RNA synthesis and measured the half-lives of the specific biliprotein transcripts under different conditions of illumination. The half-lives of the inducible PC, PE and AP transcripts were found to be 13, 25 and 20 min, respectively, and were independent of light quality. Thus the slow decline of PE transcript levels after the transfer of cultures from G to R must be due to continued synthesis of PE mRNA, rather than a consequence of differential mRNA stability under different conditions of illumination. These results suggest that the molecular basis of light regulation during complementary chromatic adaptation occurs primarily via transcriptional control. They also suggest some characteristics of the mechanisms that contribute to differential transcription of the inducible PC and PE gene sets. The kinetics of decline in the PE mRNA levels after switching from G to R favors the hypothesis that the PE gene set is controlled by a positive regulatory element; the gradual decline may result from dilution (during cell division) of a cellular factor important in promoting transcription from the PE gene set. The rapid decline in the level of inducible PC mRNA after transferring cells from R to G suggests either that a negative regulatory element is rapidly synthesized following the transfer, or that a positive regulatory element responsible for the activation of the inducible PC gene set is unstable in G.

Because R and G light pulses potentiate the synthesis of PC and PE and since the effects of these light treatments are photoreversible, it has been postulated that chromatic adaptation is controlled by a photoreceptor with action maxima in the R and G regions of the spectrum (Haury & Bogorad, 1977; Vogelmann & Scheibe, 1978). Using DNA probes derived from the inducible PC and PE gene sets, changes in transcript levels in D following R and G pulses were quantitated. Transcript levels change rapidly in response to light pulses (Oelmüller et al., 1988b). In cells acclimated to G, a 10 min R pulse yields a 5-fold increase (ca. 15% of the steady state level in R-acclimated cells) in inducible PC mRNA measured after 4 h in D. Similarly, elevated PE mRNA levels (ca. 40% of the steady state level of G- acclimated cells) are observed after an inductive G pulse. The effect of an inductive light pulse on mRNA accumulation in D can be canceled by a second pulse of inhibitory illumination.

Although photoreversible changes in abundance of PE and PC mRNAs are in some ways comparable to phytochrome-induced changes of gene expression in higher plants, the underlying mechanisms between perception of light and alteration of gene expression might be different in the cyanobacterial system. In *F. diplosiphon* responses to light pulses are immediate, persist in subsequent D, and are totally dependent on the light quality of the last irradiation. The effect of the pulse given before subjection to D can be reversed by the appropriate light treatment, even after several hours of darkness, without any decrease in the effectiveness of the second pulse. These properties of the cyanobacterial system indicate that dark reactions characteristic of phytochrome systems (dark reversion, destruction) are lacking in *F. diplosiphon*. Third, the system is not completely reversible since immediate irradiation with a second reversing light pulse does not completely negate the effect of a 10 min promoting light pulse. Thus, a fraction of the final response has aplready been triggered by the end of the 10 min irradiation. Furthermore, the change in PE and PC mRNA levels are more rapid than those observed in most phytochrome-regulated systems (Schäfer & Briggs, 1987). Finally, in the cyanobacterial system, both R and G induce both positive and negative responses. Thus, if a single photoreversible pigment system is involved, we cannot conclude which form is biologically active, or whether both are active (leading to repression in one configuration and activation in the other). For phytochrome, abundant evidence suggests that the P_{fr} form is the biologically active species.

ANALYSIS OF CYANOBACTERIAL MUTANTS

One advantage of the phycobilisome system is the ease with which pigment mutants can be obtained. Mutations with altered phycobilisome structure were of primary importance in the development of models of the architecture of these light-harvesting complexes (Yamanaka & Glazer, 1981; Lundell et al., 1981; Gingrich et al., 1982; Anderson & Eiserling, 1986). Although mutations can be induced with standard mutagenic agents like UV light and nitrosoguanidine, we have observed that stress, such as low levels of antibiotics or electroporation, can cause increased frequencies of pigment mutants in *F. diplosiphon, Synechocystis* 6701, and 6803.

This could reflect a stress-induced transposition event since insertion sequence elements have been discovered in *F. diplosiphon* (Tandeau de Marsac, pers. commun.). Some of the current research in our laboratory involves the characterization of photoregulatory and assembly-defective mutants.

Electroporation is a means of introducing DNA into cells by causing a transient perforation of the cell wall with a slow discharge of current. In experiments where ca. 90% of *F. diplosiphon* cells were killed by the electroporation, a high frequency of pigment mutants was observed among survivors. Many of these mutants were isolated and characterized for absorbance and emission properties, protein composition of isolated phycobilisomes, and the levels of specific phycobiliprotein transcripts after growth in either R or G (Bruns et al., 1988). We have isolated at least three classes of mutants with altered chromatic adaptation. The *F. diplosiphon* red pigmented (FdR) class exhibits the same pigmentation as Fd33 cells grown in G regardless of light conditions. Therefore, FdR strains cannot acclimate to R by suppressing PE transcription and activating transcription from the inducible PC gene set. *F. diplosiphon* green pigmented (FdG) mutants have little or no PE in either G or R, yet the inducible PC gene set is properly photoregulated. A third class of mutants, the *F. diplosiphon* blue pigmented (FdB) mutants, regulate the PE gene set normally, but the inducible PC gene set is constitutively expressed. The FdB phenotype is more complex than FdR or FdG phenotypes as it also appears to have increased levels of AP (3-fold), in both R and G, relative to Fd33. One consequence of this mutation is that FdB cells grown in G may have two types of rod substructures which bind to the phycobilisome cores since both inducible PC and PE gene sets are expressed under these conditions.

The FdR mutants suggest a common link in the transduction pathway for photoregulation of PE and PC since expression of both gene sets is affected in this mutant, whereas the FdG and FdB mutants suggest a branching of the regulatory pathway after which PC and PE photoregulation can be affected independently. The FdR, FdG, and FdB mutants are similar in their photoregulatory properties to mutants reported by others (Cobley & Miranda, 1983; Tandeau de Marsac, 1983). However, it is important that such mutants be completely analyzed at the molecular and structural levels before classifications are made with respect to photoregulation.

An important aspect of the analysis of photoregulatory mutants will only be realized when efficient gene transfer is established for *F. diplosiphon*. Unlike some cyanobacteria, it is not naturally competent for genetic transformation. Although conjugation by tri-parental mating is feasible for this organism (Cobley, 1985; Wolk et al., 1984), some technical problems led us to explore electroporation as a means of introducing DNA into *F. diplosiphon*. With the aid of John Cobley, (Univ. of San Francisco) we have constructed a hybrid vector that contains a pBR322 fragment with the bom and ori regions, an origin of replication from a plasmid and a kanamycin resistance marker. Using this vector we have established voltage and discharge conditions that yield a high frequency of kanamycin resistant electroporants which are currently being characterized.

We are also attempting to complement photoregulatory mutants with plasmid libraries constructed by placing Fd33 genomic DNA into a unique site in our electroporation vector. Following electroporation the photoregulatory mutants are being screened for reversion to the Fd33 phenotype. The introduced plasmid will be isolated and the cloned sequences characterized. This type of analysis may enable us to isolate genes encoding elements of the signal transduction pathway for chromatic adaptation. Such a step is crucial in the molecular description of this process and has excellent potential for success.

The unicellular *Synechocystis* 6701 is also capable of chromatic adaptation but differs from *F. diplosiphon* in that it only photoregulates the level of PE (ca. two times the amount of PE is present in G-grown relative to R-grown cells). Light quality does not affect PC levels (Tandeau de Marsac, 1977). The rod substructure of the phycobilisome from *Synechocystis* 6701 has two hexamers of PC and one, two, or no PE hexamers, depending upon the light conditions during growth. Two UV-induced mutants of *Synechocystis* 6701 with defects in phycobiliprotein assembly have been isolated and characterized (Anderson et al., 1984). The UV16 mutant has a block in PC assembly that prevents the formation of hexamers. Comparisons of PC from wild type (WT) and UV16 suggested that the α subunit of PC is the site of the assembly lesion. Strain UV16-40 was derived from UV-mutagenesis of UV16 and has about 10% of the WT level of PE, as well as the UV16 PC mutation. Comparisons of WT and UV16-40 PE proteins showed that 95% of the mutant PE$^{\beta}$ subunit was missing one of its three chromophores. PE did not assemble beyond the hexamer stage and much of the defective β subunit appeared to be proteolytically degraded, leaving some of the PE^α subunits free in the cytoplasm. The UV16 and UV16-40 strains are two examples of structural mutations that may cause aberrations in the assembly of phycobilisome components. Sequence analysis of the PE and PC gene sets from WT and UV16 or UV16-40 may allow us to define the lesions and their structural consequences, especially since any changes in the biliproteins can be precisely localized in the trimeric and hexameric structures determined by X-ray crystallography (e.g. Schirmer et al., 1986).

Recently, we have cloned and sequenced the PC gene sets from the WT and UV16 strains of *Synechocystis* 6701 and found that a WT conserved proline residue near the N-terminus of PC^α is changed to a leucine in UV16 (Anderson & Grossman, unpublished). This mutation could cause the block in PC assembly since the altered site is in a domain of the α subunit that mediates hexamer formation from two trimers. The bulky leucine is positioned where it could sterically hinder the α to α associations between trimers.

The complete UV16 phenotype is complex and may encompass secondary consequences of interrupted PC assembly (Anderson et al., 1987). The PC level in UV16 is about 35% of that in WT. The PE content is reduced to 50% of the WT level and a PE-related linker that is normally absent in R has become constitutive. Since there are no rod assembly sites for PE in UV16, the PE that is made aggregates into insoluble granules that can be seen in the fluorescence microscope. We have no genetic proof that the complete UV16 phenotype is due solely to the lesion in PC^α, although some of the traits can be interpreted as a direct consequence of blocking PC assembly.

Structural mutations, like those of UV16 and UV16-40, are not common products of random mutagenesis. However, gene transfer technology, using the PC crystal structure as a guide and *in vitro* mutagenesis, will allow us to construct specific lesions in phycobiliproteins and analyze their effects on phycobilisome structure and assembly. Chromophore binding sites can be altered to evaluate the role of chromophore attachment in assembly. Linker polypeptides can be modified to determine the domains which interact with the biliproteins and direct rod assembly and the role that glycosylation (Reithman et al., 1987) plays in phycobilisome biosynthesis and energy transfer. The consequences of eliminating structural components of the phycobilisome by targeted gene replacements may also be informative. Many of these experiments can be performed in cyanobacteria that are already transformation competent, like *Anacystis nidulans* R2 (Golden & Sherman, 1984), *Synechococcus* 7002 (Bryant et al., 1987), and *Synechocystis* 6803 (Vermaas et al., 1987). Whereas none of these organisms synthesize PE or exhibits chromatic adaptation, they can be used to address many of the basic questions concerning phycobilisome biosynthesis.

In summary, the development of molecular techniques for isolating and analyzing cyanobacterial genes coupled with the extensive structural and biochemical analyses of phycobilisomes from WT and mutant organisms, provide us with a system poised for detailed characterization at the level of biosynthesis. From such studies, we expect to learn some of the basic principles of photoperception, signal transduction and the assembly of complex cell structures.

ACKNOWLEDGMENT

We would like to thank Nancy Federspiel, David Laudenbach, Devaki Bhaya and Brigitte Bruns for critically reading this manuscript and Loretta Tayabas, Sabrina Robbins and Glenn Ford for excellent technical assistance. Work described here was supported by a Public Health Service Grant GM334336-01 from the National Institute of Health, a National Science Foundation Grant DCB-86 15606, and the Carnegie Institution of Washington. LKA was supported by a postdoctoral fellowship from the National Science Foundation DMB-8508808.

REFERENCES

Allen MM, Smith AJ (1969). Nitrogen chlorosis in blue green algae. Arch Microbiol 69:111-120.
Anderson LK, Eiserling FA (1986). Asymmetrical core structure in phycobilisomes of the cyanobacterium *Synechocystis* 6701. J Mol Biol 191:441-451.
Anderson LK, Grossman AR (1987). Phycocyanin genes in the cyanobacterium *Synechocystis* 6701 and a potential gene rearrangement in a pigment variant. In J Biggins (ed): "Progress in Photosynthesis" Proc Int Congr Photosyn Res 4:817-20.
Anderson LK, Rayner MC, Eiserling FA (1984). Ultraviolet

mutagenesis of *Synechocystis* sp. 6701. Mutations in chromatic adaptation and phycobilisome assembly. Arch Microbiol 138:237-243.

Anderson LK, Rayner MC, Eiserling FA (1987). Mutations that affect structure and assembly of light-harvesting proteins in the cyanobacterium *Synechocystis* sp. strain 6701. J Bacteriol 169:102-109.

Belknap WR, Haselkorn R (1987). Cloning and light regulation of expression of the phycocyanin operon of the cyanobacterium *Anabaena*. EMBO J 6:871-884.

Bogorad L (1975). Phycobiliproteins and complementary chromatic adaptation. Ann Rev Plant Physiol 26:369-401.

Bryant DA, Cohen-Bazire G (1981). Effects of chromatic illumination on cyanobacterial phycobilisomes. Evidence for the specific induction of a second pair of phycocyanin subunits in *Pseudanabaena* 7409 grown in red light. Eur J Biochem 119:415-424.

Bryant DA, de Lorimier R, Guglielmi G, Stirewalt VL, Dubbs JM, Illman B, Porter RD, Stevens SE Jr (1985a). Genes for phycobilisome components in *Synechococcus* 7002, *Pseudanabaena* 7409 and *Mastiglocladus laminosus*. Int Symp Photosyn Prokary 5:103.

Bryant DA, de Lorimier R, Lambert DH, Dubbs JM, Stirewalt VL, Stevens Jr SE, Porter RD, Tam J, Jay E (1985b). Molecular cloning and nucleotide sequence of the α and β subunits of allophycocyanin from the cyanelle genome of *Cyanophora paradoxa*. Proc Natl Acad Sci USA 82: 3242-3246.

Cobley JG, Miranda RD (1983). Mutations affecting chromatic adaptation in the cyanobacterium *Fremyella diplosiphon*. J Bacteriol 153:1486-1492.

Cobley JG (1985). Chromatic adaptation in *Fremyella diplosiphon*. I. Mutants hypersensitive to green light. II. Construction of mobilizable vectors lacking sites for FDI I and II. Int Symp Photosyn Prokary 5:105.

Conley PB, Lemaux PG, Grossman AR (1985). Cyanobacterial light harvesting complex subunits encoded in two red light induced transcripts. Science 213:550-553.

Conley PB, Lemaux PG, Lomax TL, Grossman AR (1986). Genes encoding major light harvesting polypeptides are clustered on the cyanobacterial genome of *Fremyella diplosiphon*. Proc Natl Acad Sci USA 83:3924-3928.

Conley PB, Lemaux PG, Grossman, AR. (1988). Molecular characterization and evolution of sequences encoding light harvesting components in the chromatically adapting cyanobacterium *Fremyella diplosiphon*. J Mol Biol 199:447-465.

de Lorimier R, Bryant DA, Porter RD, Liu W-Y, Jay E, Stevens Jr SE (1984). Genes for α and β phycocyanin. Proc Natl Acad Sci USA 81:7946-7950.

Dubbs J, Bryant DA (1987). Organization of the genes encoding phycoerythrin and the two differentially expressed phycocyanins in the cyanobacterium *Pseudanabaena* PCC 7409. In Biggins J (ed): "Progress in Photosynthesis" Proc Int Congr Photosyn Res 4:761-764.

Egelhoff T, Grossman AR (1983). Cytoplasmic and chloroplast synthesis of phycobilisome polypeptides. Proc Natl Acad Sci USA 80:3339-3343.

Füglistaller P, Suter F, Zuber H (1985). Linker polypeptides of the phycobilisome from the cyanobacterium *Mastigocladus laminosus*: amino acid sequence and relationships. Biol Chem Hoppe-Seyler 366:993-1001.

Gantt E, Conti SF (1966a). Phycobiliprotein localization in algae. Brookhaven Symp Biol 19:393.

Gantt E, Conti SF (1966b). Granules associated with the chloroplast lamellae of *Porphyridium cruentum*. J Cell Biol 29:423-430.

Gantt E, Lipschultz CA, Zilinskas B (1976). Further evidence for a phycobilisome model from selective dissociation, fluorescence emission, immunoprecipitation and electron microscopy. Biochim Biophys Acta 430:375-388.

Glazer AN (1985). Light harvesting by phycobilisomes. Annu Rev Biochem 14:47-77.

Glazer AN, Lundell DJ, Yamanaka G, Williams RC (1983). The structure of a simple phycobilisome. Ann Microbiol (Inst. Pasteur, Paris) 134B:159-180.

Glazer AN, Yeh SW, Webb SP, Clark JH (1985). Disk-to-disk transfer as the rate limiting step for energy flow in phycobilisomes. Science 227:419-423.

Golden SS, Sherman LA (1984). Optimal conditions for genetic transformation of the cyanobacterium *Anacystis nidulans* R2. J Bacteriol 158:36-42.

Grossman AR, Lemaux PG, Conley PB (1986). Regulated synthesis of phycobilisome components. Photochem Photobiol 44:827-837.

Grossman AR, Lemaux PG, Conley PB, Bruns BV, Anderson LK (1988). Characterization of phycobiliprotein and linker polypeptide genes in *Fremyella diplosiphon* and their regulated expression during complementary chromatic adaptation. Photosyn Res 17:23-56.

Haury JF, Bogorad L (1977). Action spectra for phycobili-protein synthesis in a chromatically adapting cyano-phyte. Plant Physiol 60:835-839.

Houmard J, Mazel D, Moguet C, Bryant DA, Tandeau de Marsac N (1986). Organization and nucleotide sequence of genes encoding core components of the phycobilisomes from *Synechococcus* 6301. Mol Gen Genet 205:404-410.

Lemaux PG, Grossman A (1984). Isolation and characterization of a gene for a major light-harvesting polypeptide from *Cyanophora paradoxa*. Proc Natl Acad Sci USA 81:4100-4104.

Lemaux PG, Grossman AR (1985). Major light-harvesting polypeptides encoded in polycistronic transcripts in a eukaryotic alga. EMBO J 4:1911-1919.

Lind LK, Kalla SR, Lönneborg A, Öquist G, Gustafsson P (1985). Cloning of the β phycocyanin gene from *Anacystis nidulans*. FEBS Lett 188:27-32.

Lomax TL, Conley PB, Schilling J, Grossman AR (1987). Iso-lation and characterizatio of light-regulated phycobil-isome linker polypeptide genes and their transcription as a polycistronic mRNA. J Bacteriol 169:2675-2684.

Lönneborg A, Lind LK, Kalla SR, Gustafsson P, Öquist G (1985). Acclimation processes in the light-harvesting system of the cyanobacterium *Anacystis nidulans* follow-ing a light shift from white to red. Plant Physiol 78:110-114.

Lundell DJ, Glazer AN (1983a). Molecular architecture of light-harvesting antenna. Quarternary interactions in the *Synechococcus* 6301 phycobilisome core as revealed by partial tryptic digestion and circular dichroism studies. J Biol Chem 258:8708-8713.

Lundell DJ, Glazer AN (1983b). Molecular architecture of light-harvesting antenna. Core substructure in *Synechococcus* 6301 phycobilisomes: Two new allophyco-cyanin and allophycocyanin B complexes. J. Biol Chem 258:902-908.

Lundell DJ, Glazer AN (1983c). Molecular architecture of a light-harvesting antenna. Structure of the 18S core-rod subassembly of *Synechococcus* 6301 phycobilisomes. J Biol Chem 258:894-901.

Lundell DJ, Williams RC, Glazer AN (1981). Molecular architecture of light harvesting antenna. In vitro assembly of the rod substructures of *Synechococcus* 6301 phycobilisomes. J Biol Chem 256:3580-3592.

Mazel D, Guglielmi G, Houmard J, Sidler W, Bryant DA, Tandeau de Marsac N (1986). Green light induces transcription of phycoerythrin operon in the cyanobacterium *Calothrix* 7601. Nucl Acids Res 14:8279-8290.

Mazel D, Houmard J, Tandeau de Marsac N. (1988). A multigene family in *Calothrix* sp. PCC 7601 encodes phycocyanin, the major component of the cyanobacterial light-harvesting antenna. Molec. Gen. Genet. 211:296-304.

Oelmüller R, Conley PB, Federspiel N, Briggs WR, Grossman AR (1988a). Changes in accumulation and synthesis of transcripts encoding phycobilisome components during acclimation of *Fremyella diplosiphon* to different light qualities. Plant Physiol, in press.

Oelmüller R, Grossman AR, Briggs WR (1988b). Photoreversibility of the effect of red and green light pulses on the accumulation in darkness of mRNAs coding for phycocyanin and phycoerythrin in *Fremyella diplosiphon*. Plant Physiol, in press.

Palmer J (1985). Comparative organization of chloroplast genomes. Ann Rev Genet 19:325-354.

Pilot TJ, Fox JL (1984). Cloning and sequencing of the genes encoding α and β subunits of C-phycocyanin from the cyanobacterium *Agmenellum quadruplicatum*. Proc Natl Acad Sci USA 81:6983-6987.

Reithman HC, Mawhinnet TP, Sherman LA (1987). Phycobilisome associated glycoproteins in the cyanobacterium *Anacystis nidulans* R2. FEBS Lett 215:209-214.

Schäfer E, Briggs WR (1987). Photomorphogenesis from signal perception to gene expression. Photobiochem Photobiophys 12:305-320.

Schirmer T, Huber R, Schneider M, Bode W, Miller M, Hackert M (1986). Crystal structure analysis and refinement at 2.5 Å of hexameric C-phycocyanin from the cyanobacterium *Agmenellum quadruplicatum*. The molecular model and its implication for light-harvesting. J Mol Biol 188:651-676.

Tandeau de Marsac N (1977). Occurrence and nature of chromatic adaptation in cyanobacteria. J Bacteriol 130:82-91.

Tandeau de Marsac, N (1983). Phycobilisomes and complementary chromatic adaptation in cyanobacteria. Inst Pasteur, 81:201-254.

Vermaas WFJ, Williams JGK, Chisholm DA, Arntzen CJ (1987). Site directed mutagenesis in the photosystem II gene *psbD*, encoding D2 protein. In Biggins J (ed): "Progress in Photosynthesis" Proc Int Congr Photosyn 4:761-764.

Vogelmann TC, Scheibe J (1978). Action spectrum for chromatic adaptation in the blue green alga *Fremyella diplosiphon*. Planta 143:233-239.

Wolk CP, Vonshak A, Kehoe P, Elhai J (1984). Construction of shuttle vectors capable of conjugative transfer from *Escherichia coli* to nitrogen-fixing filamentous cyanobacteria. Proc Natl Acad Sci USA 81:1561-1565.

Yamanaka G, Glazer AN (1981). Dynamic aspects of phycobilisome structure: Modulation of phycocyanin content of *Synechococcus* phycobilisomes. Arch Microbiol 130:23-30.

Algae as Experimental Systems pages 289–302
© **1989 Alan R. Liss, Inc.**

CIRCADIAN RHYTHMS IN DINOFLAGELLATES AS EXPERIMENTAL
SYSTEMS FOR INVESTIGATING CIRCADIAN TIMING

Beatrice M. Sweeney

Department of Biological Sciences, University
of California, Santa Barbara, California 93106

Many organisms of all levels of complexity are able
to tell time. The evidence for this comes from cycles in
physiological processes that are reproduced faithfully
in the absence of time information from the environment
(Sweeney, 1987). The most thoroughly studied cycles to
date are those with periods that match the 24 hour day,
the circadian rhythms. In spite of much research, the
clock mechanism that controls these cycles is still
largely unknown. Here then is a very interesting and
important problem.

For investigating the nature of circadian time-
keeping, what is needed is an organism that clearly
shows a circadian rhythm of large amplitude, one that
persists under conditions where there are no daily time-
cues. For many years I have used as experimental
material the marine planktonic dinoflagellates, in
particular Gonyaulax polyedra (Fig. 1) and more recently
Pyrocystis fusiformis (Fig. 2). Both show very distinct
periodicities in bioluminescence, photosynthesis and
cell division that continue for at least two weeks under
constant conditions, constant light (LL) or constant
darkness (DD) and constant temperature without
environmental time-cues. Figure 3 illustrates the long
persistence of the rhythm in bioluminescence in G.
polyedra. Cells were kept in continuous light after a
last night (black bar on abscissa) of the entraining   LD
cycle. My lab and I have studied circadian rhythms in
some other dinoflagellates, including Ceratium furca,
Heterocapsa (Glenodinium) pygmaea and Heterocapsa
(Cachonina) illdefina (Sweeney, 1984). Still other
dinoflagellates no doubt also have a circadian clock.

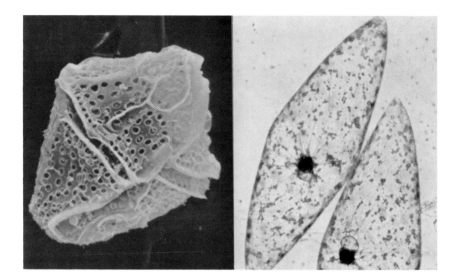

Fig. 1. (left) <u>Gonyaulax polyedra</u>, scanning electron micrograph. X 2000. (from Sweeney, 1983, Fig. 5). Fig. 2.(right)<u>Pyrocystis fusiformis</u>, light micrograph of cell during first day following division. X 140.

It is of course important that any organism chosen as an experimental subject be easy to grow in the laboratory. Fortunately the photosynthetic marine dinoflagellates can be cultured without difficulty, provided that a source of uncontaminated seawater is available. It is possible but somewhat more difficult to culture these marine organisms in a medium made up with artificial seawater. The medium used is f/2 (Guillard & Ryther, 1962), a simple formula for enriching natural seawater. For culturing dinoflagellates silicon is omitted and soil extract (0.1-0.5%) is added. The latter is not required over the short term but its inclusion improves cultures when a strain is maintained for a long time in the laboratory.

For biochemical and molecular biological studies a large amount of material is often required. The only limitation to the amount of material available at any one time is the size and number of the containers and

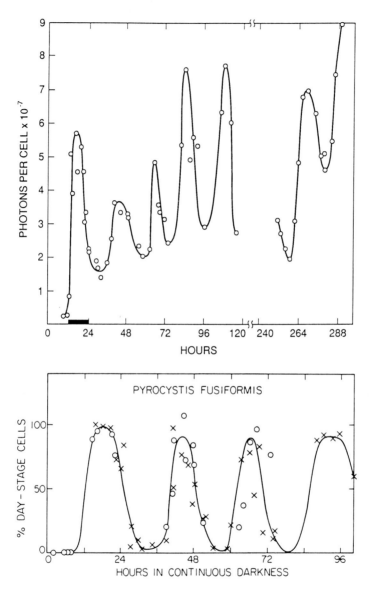

Fig. 3. (above) Circadian rhythm in bioluminescence in
Gonyaulax polyedra in LL. Fig. 4. (below) Circadian
rhythm in chloroplast migration in P. fusiformis in DD.
(from Sweeney, 1981, Fig. 6).

the difficulty of controlling the light and temperature
in large culture vessels. In general, the
dinoflagellates do not tolerate long-continued shaking
or aeration of the culture medium, so that a fairly
extensive liquid-air interface must be provided to allow
diffusion of enough CO_2 for photosynthesis. This
complicates the geometry of possible containers for mass
cultures. The smaller dinoflagellates such as H.
pygmaea, are more tolerant of shaking and have
intrinsically higher growth rates than Gonyaulax and
Pyrocystis. For these reasons, H. pygmaea has been used
for pigment studies (Boczar et al. 1980). While the
clonal cultures are routinely maintained for long times
(10-20 years), old cultures sometimes change in
physiological properties (Sweeney, 1986). For this
reason, I reisolate clonal cultures from the sea when a
desired organism is common in the plankton.

For rhythm studies of populations, the circadian
cycles of all the cells must be synchronized. This is
accomplished easily by growing cultures in a light-dark
cycle not too different from that of the day and night
of the environment. Under these conditions, circadian
rhythms become entrained to the light-dark cycle, so
that their periods exactly match that of the
environment. The timing of the entraining cycle may be
chosen for the convenience of the investigator, since
rhythms will adjust to the new light-dark cycle within
4-5 days. For example, when studying bioluminescence,
which reaches its peak during the night, the light-dark
cycle can be adjusted so that the maximum in light
emission falls during the natural day.

While entrained to a 24 h light-dark cycle, a
circadian rhythm is indistinguishable from a light-
driven response. The natural period, in fact the very
existence of a circadian rhythm, can only be measured
under constant environmental conditions. It is desirable
that the rhythm under study continue to be measurable
for at lease 3-4 days in constant conditions, not only
to confirm the presence of a circadian clock but to
allow time to measure the period and possibly to
manipulate both period and phase. Both Gonyaulax
polyedra (Fig. 3) and Pyrocystis fusiformis (Fig. 4)
fulfill this requirement. Cultures are first entrained

to a light-dark cycle and then transferred to constant
light and temperature. This transfer must take place at
a time when the change in light does not reset the phase
of the rhythm (Sweeney, 1983, 1987), usually simply by
extending the light period at the same irradiance or
introducing a different irradiance at the beginning of a
light period of the entraining cycle.

Measurements of bioluminescence are simple to make.
The light output is detected with a photomultiplier,
recorded as intensity or summed with a capacitor. The
bioluminescent flash of dinoflagellates can be induced
by mild acidification of the medium or by mechanical
stimulation. The light emitted is quite bright, about
10^8 photons/cell in G. polyedra at the maximum and a
thousand times brighter in P. fusiformis, more than 10^{11}
photons/cell. Some aspects of the rhythm in
bioluminescence do not require stimulation, for example
the frequency of spontaneous flashing that mirrors the
peak in stimulated bioluminescence and the dim glow that
occurs later in the night in Gonyaulax polyedra. Now it
is possible to computerize measurements of the
spontaneous luminescent flashing and the glow. Such a
system was built in the laboratory of J. W. Hastings
(Broda et al. 1986) and is working very well, even
plotting the rhythms as the data is taken.

The rhythm in photosynthesis may be measured either
as the evolution of oxygen detected with an oxygen-
sensitive electrode or as the uptake of radiolabelled
carbon dioxide. These techniques are well worked out and
are in common practice.

The rhythms in cell division are also easy to
measure. In Gonyaulax, recently divided cells remain as
pairs for a short time and so counts of the number of
pairs can indicate the number of cells undergoing
division. In P. fusiformis, dividing cells have a very
distinctive morphology, easily differentiated from
vegetative cells.

In P. fusiformis, the chloroplasts migrate all the
way from the tips of the cell to its center close to the
nucleus and back to the tips every day (Fig. 4). This
rhythm can be measured with only a dissecting microscope

since it is very simple to distinguish whether the
chloroplasts are dispersed or aggregated (Sweeney, 1981,
1984). The change from one state to the other only
occupies a fraction of an hour, so that the timing of
the rhythm is sharply defined.

Marked circadian rhythms that are easy to measure
in an organism that can be cultured in the laboratory
without difficulty are absolutely necessary in a good
experimental system for investigating circadian timing.
While not absolutely required, there are properties that
could make an experimental system more favorable. An
example is the presence in one organism of more than a
single rhythmic process. The timing of the different
rhythms can be compared. The simultaneous measurement of
three rhythms in G.polyedra led to the conclusion that
there is only one clock controlling timing in this
single cell, an important conclusion (McMurry and
Hastings, 1972). In G. polyedra, three somewhat
different rhythms in bioluminescence can be
distinguished, one in stimulated flashing, one in
spontaneous flashing and one in a steady glow of low
intensity. A change in the pyrenoid accompanies the
rhythm in photosynthesis (Herman and Sweeney, 1975). In
P. fusiformis, rhythms of stimulated bioluminescence and
chloroplast migration are the most distinct. Biochemical
rhythms of course must underly these overt
rhythmicities. In fact the circadian clock will also be
seen to be rhythmic when we can recognize it.

Another advantage of dinoflagellates as model
systems is the nature of the physiology that expresses
rhythmicity. Bioluminescence is already fairly well
understood at the biochemical level in a number of
organisms including dinoflagellates (Hastings, 1986).
Photosynthesis has been studied extensively and much is
known about its components and how they interact. In
animals on the other hand, the most commonly studied
circadian rhythm is that in motor activity. This is not
a single process nor one that is easy to unravel at the
biochemical level. Thus an organsism with circadian
rhythms in specific physiological processes that can be
more easily dissected is a better choice as experimental
tool.

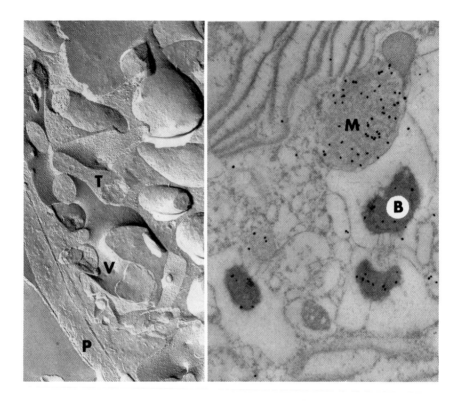

Fig. 5. (left) An electron micrograph of a freeze-
fractured Gonyaulax polyedra, Note many membrane-bound
organelles; T = trichocyst; V = vacuole.P = pellicle X
30,000. Fig. 6. (right) Bioluminescent organelles (B)
in Pyrocystis fusiformis labelled with antiluciferase
and gold particles. Note delicate strands connecting
these organelles to vacuolar membrane A similar
structure was labelled in Gonyaulax polyedra (Nicolas et
al., 1987b). (M): mucocyst> From sections prepared
during the day by M.-T. Nicolas. X 17,700.

        That dinoflagellates live as single cells
makes some studies easier than they would be in a
multicellular organism. There is no question of which
cell is responsible for timing. Fixation for
transmission electron microscopy and freezing for

freeze-fracture can be carried out with concentrated cell suspensions. Both techniques have been used successfully to visualize the cytoplasmic organization in Gonyaulax (Fig. 5).

Pyrocystis fusiformis offers several advantages over Gonyaulax. One is the ease with which organelles can be seen in living cells with the light microscope. Pyrocystis is a large cell, as much as 1 mm in length. It is thin in one dimension, much the same shape as a surf board, and contains a large vacuole, making viewing optimal. Gonyaulax on the other hand is much smaller, spherical and filled with opaque organelles. Circadian changes in the chloroplasts of Gonyaulax must be observed in fixed cells using the electron microscope (Herman & Sweeney, 1975), while migration of the chloroplasts in Pyrocystis can be detected in living cells at low magnification.

The optical properties of Pyrocystis cells make it possible to see the microsources of bioluminescence under the microscope as glowing bodies during the day phase and as flashing spots at night (Sweeney, 1982a; Widder & Case, 1981). Sections stained with an antibody to luciferase and a second antibody labelled with gold particles have helped to identify the microsources of bioluminescence in both Gonyaulax (Nicolas et al. 1987b) and Pyrocystis (Fig. 6; Nicolas et al. 1987a).

In the fluorescence microscope dinoflagellate chloroplasts can be seen easily by the autofluorescence of chlorophyll. Cells may be stained with DNA-specific dyes such as DAPI or antibodies conjugated to fluorescent molecules. With DAPI-stained cells of P. fusiformis, the nucleus can be observed (Sweeney, 1982b), making possible the study of the unusual nuclear division, (Fig. 7). We have also used antibodies against tubulin from chicken brain to visualize microtubules during the migration of the chloroplasts in P.fusiformis (Fabros & Sweeney, unpublished).

The bioluminescence of single cells can be measured both in Gonyaulax (Krasnov et al. 1981) and in

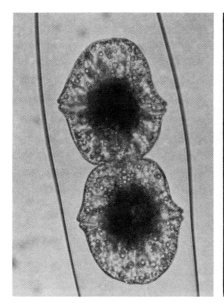

Fig. 7. Dividing cells of Pyrocystis fusiformis: a.
light micrograph of pair of cells; note girdle in each
new cell and chloroplasts close to nucleus X 1000. b.
Slightly earlier dividing pair stained with DAPI showing
strands of DNA pulling apart without conventional
spindle: chloroplast DNA visible as diffuse bright spots
X 1000.

Pyrocystis (Widder & Case, 1981; Sweeney, 1982b). The
photosynthesis of a single Gonyaulax has been measured
using a Cartesian reference diver apparatus (Sweeney,
1960).

     At present, only two properties that are certainly
belonging to the circadian clock, as distinct from the
rhythms that it controls: the period and the phase. It
is reasonably certain that the rhythmic processes,
bioluminescence, photosynthesis and cell division, are
not an intrinsic part of the clock machinery, since each
can be altered by inhibitors or other means without
affecting timing. Two kinds of approaches can be used to
investigate the nature of the circadian clock. One is to

try to alter timing directly. This method has been widely used with Gonyaulax. The strategy is to treat cultures with a biologically active substance of which the target in other cells is known and to look for either phase shifts following exposure to the drug shorter than one circadian period or to leave the drug in contact with the cells and look for changes in period (Sweeney, 1983). Such experiments using the inhibitors of protein synthesis on 80s ribosomes, cycloheximide and anisomycin, have offered evidence that the clock requires protein synthesis on cytoplasmic ribosomes (Walz & Sweeney, 1979; Dunlap et al. 1980; Rensing et al. 1980; Taylor et al. 1982). The success of this method depends on the specificity of the drugs being the same in dinoflagellates as in other organisms.

The other approach is to trace the transduction in reverse from the overt rhythm toward the controlling clock. This last plan of action has led to the discovery of a number of biochemical rhythms, some of which are clearly part of a known rhythm and others which are not so obviously connected. An example of the first type is the rhythm in the amount of luciferase (Dunlap & Hastings, 1980; Johnson et al. 1984) and luciferin-binding protein. At present, it appears that the luciferase rhythm is controlled at translation since the amount of mRNA for luciferase has not been found to change during the circadian cycle (Hastings, personal communication). There is, however, a rhythm in total RNA and in some specific RNAs (Walz et al., 1983), that is of the second type, so far not connected to any overt circadian rhythm in Gonyaulax. The accessibility of the photosynthetic machinery has made it possible to localize rhythmicity to PSII (Samuelsson et al. 1983). Studies of the kinetics of autofluorescence (Sweeney et al. 1979) suggested an underlying rhythm in dark decay processes. Further localization of the control point in photosynthetic electron flow should be possible.

Recently the new and exciting techniques of molecular biology have been applied to the circadian clock problem. The genetics of dinoflagellates is not yet well known. This is especially true of photosynthetic species. Mutants are not available as they are in such rhythmic organisms as Drosophila and

Neurospora, principally because sexual stages have not proven easy to induce and crosses have not been commonly possible. This has been a distinct disadvantage in dinoflagellates as experimental objects in the study of circadian rhythmicity. Now, however, the new techniques of molecular biology have provided a way around this stumbling block. For example, we have been using as a probe a plasmid containing the period gene from Drosophila melanogaster (Bargiello et al. 1984), mutants of which have altered periods in several circadian rhythms, including eclosion of the adult from the puppa, activity of the adult and the courtship song of the male. This plasmid was generously supplied to us by Rob Jackson (Jackson et al., 1986). With this probe we have shown that the DNA isolated from Gonyaulax polyedra contains base sequences similar enough to the per gene of Drosophila to hydridize with it under fairly stringent conditions. We have cloned this sequence from a cDNA library of Gonyaulax, made in lambda gt10, and are currently sequencing this clone. When the sequence is available, it can be compared with with that of the per gene in the fruit fly. Next my lab and I would like to study genomic clones from Gonyaulax . If genomic sequence with similarity to our cDNA clone can be isoltaed, we will be in a position to detect control regions which may be at the 5' end of the gene and be able to find out by in situ mutagenesis how the clock gene influences the period length. This will require being able to introduce DNA into living cells of dinoflagellates. At present this has never been achieved, but new methods are rapidly becoming available.

REFERENCES

Bargiello T, Jackson R, Young M (1984). Restoration of circadian behavioral rhythms by gene transfer in Drsophila. Nature (Lond) 312:752-754.
Boczar BA, Prezelin BB, Markwell JP, Thornber JP (1980). chlorophyll c-containing pigment-protein complex from the marine dinoflagellate, Glenodinium sp. FEBS 120:243-247.
Broda H, Gooch VD, Taylor W, Aiuto N, Hastings JW (1986). Acquisition of circadian bioluminescence data

in Gonyaulax and an effect of the measurement
procedure on the period of the rhythm. J Biological
Rhythms 1:251-263.
Dunlap JC, Taylor W, Hastings JW (1980). The effect of
protein synthesis inhibitors on the Gonyaulax clock.
I. Phase-shifting effects of cycloheximide. J Comp
Physiol 138:1-8.
Dunlap JC, Hastings JW (1981). The biological clock in
Gonyaulax controls luciferase activity by regulating
turnover. J Biol Chem 141:1269-1270.
Guillard RRL, Ryther JH (1962). Studies of marine
plankton diatoms. I. Cyclotella nana Hustedt and
Detonula confervacea (Cleve) Gran. Can J Microbiol
8:229-239.
Hastings JW (1986) Bioluminescence in bacteria and
dinoflagellates. In Amesz, J, Govindjee & Fork, DC
(eds.): "Light Emission in Plants and Bacteria."
Academic Press, NY pp 363-398.
Herman EM, Sweeney BM (1975). Circadian rhythm of
chloroplast ultrastructure in Gonyaulax polyedra,
concentric organization around a central cluster of
ribosomes. J Ultrastruct Res 50:347-354.
Jackson FR, Bargiello TA, Yun S-H, Young MW (1986).
Product of the per locus of Drosophila shares
homology with proteoglycans. Nature (Lond) 320:185-
188.
Johnson CH, Roeber JF, Hastings JW (1984). Circadian
changes in enzyme concentration account for rhythm of
enzyme activity in Gonyaulax. Science 223:1428-1430.
Krasnow R, Dunlap J, Taylor W, Hastings JW (1981).
Measurements of Gonyaulax bioluminescence, including
that of single cells. In K H Nealson (ed):
"Bioluminescence, Current Prospectives". Minneapolis:
Burgess Publishing Co. pp. 52-63.
McMurry L, Hastings JW (1972). No desynchronization
among four circadian rhythms in the unicellular alga,
Gonyaulax polyedra. Science 175:1137-1139.
Nicolas M-T, Sweeney BM, Hastings JW (1987a). The
ultrastructural localization of luciferase in three
bioluminescent dinoflagellates, two species of
Pyrocystis, and Noctiluca, using antiluciferase
and immunogold labelling. J Cell Sci 87:189-196.
Nicolas MT, Nicolas G, Johnson CH, Bassot J-M, Hastings
JW (1987b). Characterization of the bioluminescent
organelles in Gonyaulax polyedra (Dinoflagellates)

after fast-freeze fixation and antiluciferase immunogold staining. J Cell Biol 105:723-735.

Rensing L, Taylor W, Dunlap J, Hastings JW (1980). The effects of protein synthesis inhibitors on the Gonyaulax clock. II. The effect of cycloheximide on ultrastructural parameters. J Comp Physiol 138:9-18.

Samuelsson G, Sweeney BM, Matlick HA, Prezelin BB (1983). Changes in photosystem II account for the circadian rhythm in photosynthesis in Gonyaulax polyedra. Plant Physiol 73:329-331.

Sweeney BM (1960). The photosynthetic rhythm in single cells of Gonyaulax polyedra. Cold Spring Harbor Symp Quant Biol 25:145-148.

Sweeney BM (1981). The circadian rhythms in bioluminescence, photosynthesis and organellar movement in the large dinoflagellate, Pyrocyatis fusiformis. In Schweiger H-G (ed): "International Cell Biology 1980-1981." Berlin: Springer Verlag pp. 807-814.

Sweeney BM (1982a). Microsources of bioluminescence in Pyrocystis fusiformis (Pyrrophyta). J Phycol 18:412-416.

Sweeney BM (1982b). Interaction of the circadian cycle with the cell cycle in Pyrocystis fusiformis. Plant Physiol 70:272-276.

Sweeney BM (1983). Circadian time-keeping in eukaryotic cells, models and experiments. Prog Phycol Res 2:189-225.

Sweeney BM (1984). Circadian rhythmicity in dinoflagellates. In DL Spector (ed): "Dinoflagellates", New York: Academic Press pp. 343-364.

Sweeney BM (1986). The loss of the circadian rhythm in photosynthesis in an old strain of Gonyaulax polyedra. Plant Physiol 80:978-981.

Sweeney BM (1987). "Rhythmic Phenomena in Plants", 2nd Ed, New York: Academic Press 172 pp.

Sweeney BM, Prezelin BB, Wong D, Govindjee (1979). In vivo chlorophyll a fluorescence transients and the circadian rhythm of photosynthesis in Gonyaulax polyedra. Photochem Photobiol 30:309-311.

Taylor WR, Dunlap JC, Hastings JW (1982). Inhibitors of protein synthesis on 80s ribosomes phase shift the Gonyaulax clock. J Exp Biol 97:121-136.

Walz B, Sweeney BM (1979). Kinetics of the

cycloheximide-induced phase changes in the biological clock in <u>Gonyaulax.</u> Proc Natl Acad Sci USA 76: 6443-6447.

Walz B, Walz A, Sweeney BM (1983). A circadian rhythm in RNA in the dinoflagellate, <u>Gonyaulax</u> <u>polyedra.</u> J Comp Physiol 151:207-213.

Widder EA. Case JF (1981). Two flash forms in the bioluminescent dinoflagellate, <u>Pyrocystis</u> <u>fusiformis.</u> J Comp Physiol 143:43-52.

POSTER PRESENTATIONS

[Address of senior author in "List of Participants" unless noted otherwise]

Allen, N.S. & S. A. O'Conner

D-myo-inositol 1,4,5-trisphosphate Transiently Inhibits Intracellular Particle Motions in Acetabularia.

Apt, K. & A. Gibor

Induction of Gall and Tumorous Growths on Marine Macroalgae by Microorganisms.

Ambrust, E.V., S. Krolikowski, S.W. Chisholm, & R.J. Olson

Induction of Spermatogenesis in a Centric Marine Diatom.

Bloodgood, R.A. & N.L Salomonsky

Flagellar Glycoprotein Dynamics as the Basis for Gliding Motility in Chlamydomonas reinhardtii.

Boczar, B.A., M.S. Shivji, T. Delaney, N. Li, & R.A. Cattolico
Structure, Expression, and Evolution of Chloroplast Genes of Algae.

Brown, R.C. & B.E. Lemmon

Evidence for Cytoplasmic Domains in Higher Plants: Supernumerary Nuclei Claim Cytoplasm in Coenocytic Microsporocytes.

Evidence for Cytoplasmic Domains in Higher Plants: The Cytoskeleton in Cytokinesis of Coenocytic Sporocytes.

Chang, J. & E.J. Carpenter

Estimating Cell Cycle Phase Durations in Phytoplankton and Its Application to Growth Rate Measurement.

Coleman, A.W.

The Ochromonas Solution to the Problem of Plastid
Continuity.

Denning, G.M. & A.B. Fulton

Isolation, Characterization and Function of Clathrin-
Coated Vesicles from Chlamydomonas.

Domozych, D.S.

The Role of the Endomembrane System During Phycoplast-
mediated Cytokinesis in Green Algal Flagellates.

Duke, C.S., A. Cezeaux & M.M. Allen

Nitrogen Starvation-induced Changes in the Polypeptide
Composition of Phycobilisomes of Synechocystis 6308.

Ehara, T., T. Osafune, S. Sumida & E. Hase

Light-independent and Dependent Processes of the Formation
of Thylakoids and Pyrenoid in Dark-grown Cells of Euglena.

Farmer, M.A. & R.E. Triemer

Flagellar Replication in Euglenoid Flagellates.

Freshwater, D.W. & D.F. Kapraun

Strain Selection and Spheroplast Propagation of Porphyra
carolinensis Coll et Cox.

Friedman, W.E.

Morphogenesis in the Seed Plant Male Gametophyte of Ginkgo
biloba.

Fritz, L., P. Milos, D. Morse & J.W. Hastings

In-situ Hybridization of Luciferin Binding Protein mRNA to
Thin-Sections of the Bioluminescent Dinoflagellate
Gonyaulax.

Fujita, R.M. & P.A. Wheeler

Metabolic Regulation of Ammonium Uptake by <u>Ulva rigida</u> (Chlorophyta).

Goff, L.J. & G. Zuccarello

Horizontal Gene Transfer in Red Algal Parasitism.

Gross, W.

Microbodies from the Xanthophyte <u>Vaucheria</u>.

Hamada, J. & T. Bando

Diploid DNA Content in Vegetative Cells of <u>Closterium ehrenbergii</u> (Chlorophyta) and the Effects of Radiations.

Harper, J.D.I. & P.C.L. John

Genetical Analysis of the <u>Chlamydomonas</u> Cell Division Cycle.

Henry, E.C. & S. Lanka

A Stable Virus Infection in a Brown Alga (Phaeophyceae).

Hill, G.J.C., M.R. Cunningham, M.M. Byrne & T.P. Ferry

Chemical Control of Androspore Morphogenesis in <u>Oedogonium donnellii</u>.

Holmes, J.A. & S.K. Dutcher

Cellular Asymmetry of <u>Chlamydomonas reinhardtii</u>.

Hotchkiss, A.T., Jr., M.R. Gretz & R.M. Brown, Jr.

The Cell Wall Composition of <u>Mougeotia</u> (Charophyceae).

Husic, H. David & Kristin Morris

Localization and Properties of Carbonic Anhydrase in <u>Chlamydomonas reinhardtii</u> and <u>Chlamydomonas moewusii</u>.

Ishida, M.R.

Phylogeny of the Nuclear and Chloroplast Genomes in Algae Based on Their DNA Base Composition and Gene Organization.

The Isotope Effects of Heavy Water on the Growth and Protein Synthesis of Unicellular Green Alga, Chlamydomonas.

Ishida, M.R. & J. Hamada

Distribution Mode of Chloroplast Nucleoids in Some Algae: An Epifluorescence Microscopy with DAPI Staining.

Jackson, S.L. & I.B. Heath

Effects of Exogenous Calcium Ions on Tip Growth, Intracellular Ca^{2+} Concentration and Actin Arrays in Fungal Hyphae.

Jacobshagen, S., B. Pelzer-Reith, W. Ried, B. Müller & C. Schnarrenberger [see Schnarrenberger]

Aldolase - A Model for the Evolution of an Enzyme in Algae.

Kapraun, D.F. & D.W. Freshwater

Cell Isolation, Parasexual Hybridization and Strain Selection in Ulvaria oxysperma (Kuetz.) Blid. for Mariculture.

Kitayama, M. Kitayama & R.K. Togasaki [see Togasaki]

Immunological Analysis of Carbonic Anhydrase in Chlamydomonas reinhardtii.

LaRoche, J. & P.G. Falkowski

Molecular Biology of Photoadaptation in Dunaliella tertiolecta.

Leizerovich, I. & M. Galun [see Galun]

Polyphenols, and not Lectins, are Responsible for Hemagglutinating Activity in Extracts of Azolla filiculoides Lam.

Liddle, L.B.

Nuclear Genome Size During the Life History of <u>Cymopolia</u> (Chlorophyta, Dasycladales).

McAuley, P.J.

Isolation of Viable Uncontaminated Algae from the Green Hydra Symbiosis.

McNaughton, E.E. & Goff, L.J.

Cytoplasmic Events Following Wounding in the Coenocytic Green Alga <u>Valoniopsis pachynema</u> (Mart.) Boergesen.

Miller, M.W.

Growth Stimulation of <u>Cyclotella cryptica</u> (Bacillariophyceae) by Two Chemical Forms of Gibberellin, GA_9 and GA_{13}.

Mitman, G.G.

A Karyological Study of <u>Porphyra umbilicalis</u> (L.) J. Agardh from Nova Scotia, Canada.

Nicolas, M.T. & J.M. Bassot

Immunocytochemical Identification of the Microsources of Bioluminescence in Dinoflagellates after Fast-freeze Fixation.

Ott, D.W.

Video Enhanced Microscopy of Organellar Streaming in <u>Vaucheria longicaulis</u>.

Owens, T.G. & E.R. Wold

Light-harvesting Antenna Organization in Photosynthetic Mutants of <u>Chlamydomonas reinhardtii</u>.

Parthasarathy, M.V. & A. Witztum

Organization of Actin and Organelle Movement in the Alga <u>Closterium</u>.

Puiseux-Dao, S., C. Karez, D. Allemand, G. deRenzis, M. Gnassia-Barelli, M. Roméo & P. Payan

Calcium Transport in Phytoplankton Marine Algae.

Rausch, H., S. Goetinck, D. Kirk, & R. Schmitt [see Goetinck]

A Molecular Phylogeny of the Order Volvocales.

Richmond, P.A., S. Lukens & C. Ericsson

A Comparison of Microtubule Arrangements in Young and Mature Nitella Internodes.

Roberts, K.R., R.M. Schneider, J.E. Lemoine & M.A. Farmer

The Microtubular Cytoskeleton of Dinoflagellates.

Romanovicz, D.K.

Production of Cellulosic Scales in the Golgi Apparatus of Pleurochrysis.

Rothschild, L.J.

Physiology of a Microbial Mat Community as a Model System for Life on Mars.

Sanders, M.A. & J.L. Salisbury

Flagellar Excision in Chlamydomonas reinhardtii: The Transition Zone.

Sandgren, C.D. & S.B. Barlow

New Approaches to Experimental Investigations of Silicon Biomineralization, Siliceous Scale Morphogenesis, and Scale Layer Assembly Using the Chrysophycean Flagellate Synura petersenii.

Sang, J. & A.W. Coleman

Two Types of Plastids Exist in the Somatic Hybrids of Bryopsis maxima.

Schreurs, W.J.A., R.L. Harold & F.M. Harold

Chemotropism and Branching in Achyla bisexualis.

Scott, J., L. Patrone & K. Klepacki

Rhodella reticulata and Rhodella violacea: Unlikely
Congeners.

Segaar, P.J.

Visualization of Complex Microtubule Systems in the Cell
Cycle of Green Algae.

Selvin, R.C. & R.R.L. Guillard

The Provasoli-Guillard Center for Culture of Marine
Phytoplankton (CCMP), a Resource for Cell Biologists.

Sogin, Mitchell L.

The Characterization of Enzymatically Amplified Eukaryotic
16S-like rDNA Coding Regions and the Inference of
Molecular Phylogeneies from Ribosomal RNA Sequences.

Sriharan, S., T.P. Sriharan & D. Bagga

Environmental Control of Lipid Production in Microalgae,
Monoraphidium minutum, Cyclotella cryptica, and Hantzschia
DI-60.

Sylvester, A.W. & J.E. Carrier

Analysis of the Mechanism of Cytoplasmic Segregation in
the Red Alga Anotrichium tenue.

Troxell, C.L.

Ionic Currents During Primary Cell Wall Morphogenesis in
Closterium and Micrasterias.

Waffenschmidt, S., R. Spessert & L. Jaenicke

The Vegetative Autolysin of Chlamydomonas reinhardtii is a
Stage Specific Enzyme.

Yentsch, C.M. & F.C. Mague

Single-laser Three-color Flow Cytometric Measurements of
Phytoplankton Viability.

AMERICAN SOCIETY FOR CELL BIOLOGY SUMMER CONFERENCE
Airlie, Virginia, June 25-29, 1988

LIST OF PARTICIPANTS

Adair, Dr. W. Steven: Department of Anatomy & Cell Biology, Tufts Medical School, Boston, MA 02111

Allen, Nina S.: Department of Biology, Wake Forest University, Reynolds Station Box 7325, Winston-Salem, NC 27109

Apt, Kirk: Department of Biological Sciences, University of California, Santa Barbara, CA 93106

Ambrust, E. Virginia: Joint Program in Biological Oceanography, MIT 48-230, Cambridge, MA 02139

Bader, Artrice, V.: DHHS-NIH, NIGMS, Westwood Bldg Room 906, 5333 Westbard Ave, Bethesda, MD 20892

Bando, Tadashi: Botanical Institute, Hiroshima University, 1-1-89 Higashisendamachi, Naka-ku, Hiroshima City, 730 Japan

Barlow, Steven: Department of Pharmacology, UTHSC Medical School, PO Box 20708, Houston, TX 77225

Bassot, Jean-Marie: Laboratoire de Bioluminescence, CNRS, 105 Boulevard Raspail, F-75006 Paris, France

Biebel, Paul: Department of Biology, Dickinson College, Carlisle, PA 17013

Bloodgood, Robert A.: Department of Anatomy & Cell Biology, University of Virginia Medical Center Box 439, Charlottesville, VA 22908

Boczar, Barbara A.: Department of Botany KB-15, University of Washington, Seattle, WA 98195

Brown, Roy C.: Department of Botany, Miami University, Oxford, OH 45056

Carrier, John: Friday Harbor Labs, University of Washington, 620 University Rd, Friday Harbor, WA 98250

Carty, Susan: Department of Biology, North Carolina A & T State University, Greensboro, NC 27411

Chang, Jeng: Marine Sciences Research Center, State University of New York, Stony Brook, NY 11794

Chopra, B.K.: Department of Biology, Johnson C. Smith University, 100 Beatties Ford Rd, Charlotte, NC 28216

Clutter, Mary E.: National Science Foundation, 1800 G Street NW, Washington, DC 20050

Coleman, Annette W.: Department of Biology & Medicine, Brown University, Providence, RI 02912

Cox, Kathleen H.: Department of Biology, University of California, Los Angeles, South Pasadena, CA 91030

Denning, Gerene M.: Department of Biochemistry, University of Iowa, 4-711 Bowen Science Bldg, Iowa City, IA 52242

Domozych, David S.: Department of Biology, Skidmore College, North Broadway, Saratoga Springs, NY 12866

Duke, Clifford S.: Department of Biological Science, Wellesley College, Wellesley, MA 02181

Dutcher, Susan K.: Department of Molecular, Cellular & Developmental Biology, University of Colorado, Boulder, Boulder, CO 80309-0347

Dyck, Lawrence A.: Department of Biological Science, Clemson University, Clemson, SC 29634-1903

Edgerton, Michael D.: Department of Biology, University of North Carolina, 010A Coker Hall, Chapel Hill, NC 27514

Ehara, Tomoko: Department of Microbiology, Tokyo Medical College 6-1-1 Shinjuka, Tokyo 160, Japan

Farmer, Mark A.: Department of Biology, University of Southwestern Louisiana Lafayette, LA 70504-2451

Faust, Maria A.: Department of Botany, Smithsonian Institution MS Center, 4201 Silver Hill Rd, Suitland, MD 20560

Foster, Kenneth: Department of Physics, Syracuse University, Syracuse, NY 13244-1130

Freshwater, Wilson D.: Department of Biological Science, University of North Carolina-Wilmington, 601 S College Rd, Wilmington, NC 28403-3297

Friedman, William E.: Department of Botany, University of Georgia, Athens, GA 30602

Fritz, Lawrence: Department of Cell and Developmental Biology, Harvard University, 16 Divinity Ave, Cambridge, MA 02138

Fujita, Rodney M.: Marine Botany-Science Division, Harbor Branch Oceanographic Institution, 5600 Old Dixie Hwy, Fort Pierce, FL 33450

Fukuda, Ikujiro: Department of Biological Science, University of Tokyo, Kagurazaka 1-3, Shijuku-ku, Tokyo, 162 Japan

Galun, Margalith: Department of Botany, Tel Aviv University, Symbiosis Research Laboratory, Tel Aviv, Israel 69978

Gantt, Elisabeth: Smithsonian/Botany Department, University of Maryland, College Park, MD 20742

Gibbs, Sarah P.: Department of Biology, McGill University, 1205 Avenue Docteur Penfield, Montreal, PQ Canada H3A 1B1

Gibor, Aharon: Department of Biological Science, University of California, Santa Barbara, CA 93106

Goetinck, Susan D.: Department of Biology, Washington University, Box 1137, St. Louis, MO 63130

Goff, Lynda J.: Department of Biology & Marine Biology, University of California, Santa Cruz, CA 95064

Goodenough, Ursula W.: Department of Biology, Box 1137, Washington University, St. Louis, MO 63130

Gretz, Michael R.: Department of Biology, George Mason University, 4400 University Dr, Fairfax, VA 22030

Gross, Wolfgang: Department of Biology, University of California at Santa Cruz, Steinhart–May St, Santa Cruz, CA 95064

Grossman, Arthur R.: Department of Plant Biology, Carnegie Institute of Washington, 290 Panama St, Stanford, CA 94305

Hamada, Jin: Department of Community Medicine, Toyama Medical & Pharmacological University, 2630 Sugitani, Toyama-shi, Toyama, 930-01 Japan

Harper, John D.I.: Plant Cell Biology Group, RSBS Australian National University, GPO Box 475, Canberra, ACT, Australia 2601

Heimke, John W.: Department of Biology, Russell Sage College, 45 Ferry St, Troy, NY 12180

Henry, Eric C.: c/o AG Muller Fakultät für Biologie, Universität Konstanz, Postfach 5560, D-7750 Konstanz, West Germany

Hill, Gerry J.C.: Department of Biology, Carleton College, One N College St, Northfield, MN 55057

Holmes, Jeffrey A.: MCD Biology, University of Colorado, Boulder, CO 80309-0347

Hotchkiss, Arland T.: Department of Plant Science, USDA-ARS, ERRC, 600 E Mermaid Ln, Philadelphia, PA 19118

Husic, H. David: Department of Chemistry, Lafayette College, Easton, PA 18042

Ishida, Masahiro R.: Department of Nuclear Biology, Research Reactor Institute, Kyoto University, Kumatori-cho, Sennangun, Osaka, 590-04 Japan

Jackson, Sandra L.: Department of Biology, York University, 4700 Keele St, North York, Ontario Canada M3J 1P3

Jaenicke, Lothar: Institut für Biochemie, Universität zu Koln, An der Bottmühle 2, D-5000 Koln 1, West Germany

Kapraun, Donald F.: Department of Biological Science, University of North Carolina-Wilmington, 601 S College Rd, Wilmington, NC 28403-3297

Koutoulis, Anthony: School of Botany, University of Melbourne, Grattan St, Parkville, Victoria Australia 3052

Kropf, Darryl L.: Department of Biology, University of Utah, Salt Lake City, UT 84112

LaClaire, John W. II: Department of Botany, University of Texas at Austin, Austin, TX 78713-7640

LaRoche, Julie: Department of Applied Science, OSD, Brookhaven National Laboratory Bldg 318, Upton, NY 11973

Lemmon, Betty E.: Department of Biology, University of Southwestern Louisiana, Lafayette, LA 70504-2451

Liddle, Larry B.: Division of Natural Science, Long Island University, Southampton, NY 11968

Ludwig, Martha: Department of Biology, McGill University, 1205 Avenue Docteur Penfield, Montreal, PQ Canada H3A 1B1

MacKay, Ron M.: Department of Marine BioScience, Atlantic Research Laboratory, 1411 Oxford St, Halifax, Nova Scotia Canada N3H 3Z1

McAuley, Paul J.: Department of Plant Science, Oxford University, Agricultural Science Bldg, Parks Rd, Oxford OX1 3P, UK

McDonald, Kent L.: Department of Molecular, Cellular & Developmental Biology, University of Colorado, Boulder, CO 80309-0347

McNaughton, Eugenia E.: Department of Biology, University of California, Santa Cruz, CA 95064

Menzel, Diedrik: Max-Planck Institut für Zellbiologie, Rosenhof, D-6802 Ladenburg, West Germany

Miller, Kenneth R.: Division of Biology & Medicine, Brown University, Providence, RI 02912

Miller, Margaret W.: Department of Biological Science, University of South Alabama, 307 University Blvd, Mobile, AL 36688

Millington, W.F.: Department of Biology, Marquette University, Milwaukee, WI 53233

Mitman, Grant G.: Department of Biology, Dalhousie University & Atlantic Research Laboratory, 1411 Oxford St, Halifax, Nova Scotia Canada B3H 1Z3

Möller, Dieter B.: Fakultät für Biologie, Universität Konstanz, Postfach 5560, D-7750 Konstanz, West Germany

Murray, Steven N.: Department of Biological Science, California State University, Fullerton, CA 92634

Nicolas, Marie-Therese: Laboratoire de Bioluminescence, CNRS, 105 Boulevard Raspail, F-75006 Paris, France

Osafune, Tetsuaki: Department of Microbiology, Tokyo Medical College, 6-6-6 Shinjuku, Tokyo, 160 Japan

Ott, Donald W.: Department of Biology, University of Akron, Buchtel Ave, Akron OH 44325

Owens, Thomas G.: Section of Plant Biology, Cornell University, Plant Science Bldg, Ithaca, NY 14853-5908

Parthasarathy, Mandayam V.: Division of Cellular Bioscience, The National Science Foundation, 1800 G St NW, Washington, DC 20550

Pickett-Heaps, Jeremy D.: MCD Biology, University of Colorado, Boulder, CO 80309-0347

Polne-Fuller, Miriam: Department of Biological Science, University of California, Santa Barbara, CA 93106

Pueschel, Curt: Department of Biology, State University of New York, Binghamton, NY 13901

Puiseux-Dao, Simone: INSERM Unité 303 "Mer & Sante", BP 3 - La Darse, F-06230 Villefranche-sur-mer, France

Quatrano, Ralph, Plant Science, Dupont Company, Experimental Station Bldg 402, Wilmington, DE 19898

Richmond, Paul A.: Department o Biological Science, University of the Pacific, Stockton, CA 95211

Roberts, Keith R.: Department of Biology, University of Southwestern Louisiana, PO Box 42451, Lafayette, LA 70504

Rodewald, Richard D.: Department of Biology, University of Virginia, Gilmer Hall, Charlottesville, VA 22901

Romanovicz, Dwight K.: Department of Biology, West Georgia College, Carrollton, GA 30118

Rothschild, Lynn J.: NASA-Ames Research Center, Mailstop 239-12. Moffett Field, CA 94035

Salisbury, Jeffrey L.: Program for Cell Biology, Center for NeuroSciences, 2119 Abington Road, Case Western Reserve University, Cleveland, OH 44106

Sanders, Mark A.: Program for Cell Biology, Center for NeuroSciences, School of Medicine, Case Western Reserve University, Cleveland, OH 44106

Sandgren, Craig D.: Department of Biological Science, University of Wisconsin-Milwaukee, PO Box 413, Milwaukee, WI 43201

Sang, Junsheng: Division of Biology & Medicine, Brown University, Providence, RI 02912

Schenkman, Rocilda P.F.: Haskins Laboratory, Pace University, 41 Park Row, New York, NY 10038

Schnarrenberger, Claus: Institute of Plant Physiology, Cell Biology/Microbiology, Free University of Berlin, Konigin-Wise-Str 12-16a, D-1000 Berlin 33, West Germany

Schneider, Robin M.: Department of Biology, University of Southwestern Louisiana, PO Box 42451, Lafayette, LA 70504

Shreurs, Willie J.A.: Division of Molecular & Cell Biology, National Jewish Center for Immunology & Respiratory Medicine, 1400 Jackson St, Denver, CO 80206

Schumm, Joan C.: Department of Biology, Wake Forest University, Winston-Salem, NC 27109

Scott, Joseph L.: Department of Biology, College of William & Mary, Williamsburg, VA 23185

Scott, O. Tacheeni: Department of Biology, Northern Arizona University, NAU Box 5640, Flagstaff, AZ 86011

Segaar, Peter J.: Department of Cell Biology & Genetics, State University of Leiden, Schelpenkade 6, PO Box 9614, NS-2300 RA Leiden, The Netherlands

Selvin, Rhonda C.: CCMP, Bigelow Laboratory for Ocean Science, McKown Pt, West Boothbay Harbor, ME 04575

Shevlin, Dennis, E.: Department of Biology, Trenton STate College, Hillwood Lakes, CN 4700, Trenton, NJ 08650-4700

Sogin, Mitchell L.: Division of Molecular & Cell Biology, National Jewish Center for Immunology & Respiratory Medicine, 1400 Jackson St., Denver, CO 80206

Sriharan, Shobha: Division of Natural & Applied Science, Selma University, Selma, AL 36701

Sriharan, T.P.: Division of Natural & Applied Science, Selma University, Selma, AL 36701

Starr, Richard C.: Department of Botany, University of Texas at Austin, Austin, TX 78713-7640

Stein-Taylor, Janet R.: PO Box 371, Glencoe, IL 60022

Sweeney, Beatrice M.: Department of Biological Sciences, University of California, Santa Barbara, Santa Barbara, CA 93106

Sylvester, Anne W.: Department of Biological Science, Stanford University, Stanford, CA 94305

To, LeLeng: Department of Biological Science, Goucher College, Dulaney Valley Rd, Towson, MD 21204

Togasaki, Robert K.: Department of Biology, Indiana University, Bloomington, IN 47405

Troxell, Cynthia L.: Division of Molecular & Cell Biology, National Jewish Center for Immunology & Respiratory Medicine, 1400 Jackson St., Denver, CO 80206

van den Ende, Hermann: Department of Molecular Cell Biology, University of Amsterdam, Kruislaan 318, NS-1098 SM Amsterdam, The Netherlands

Waaland, Susan D.: Department of Biology, University of Puget Sound, Tacoma, WA 98416

Waffenschmidt, Sabine: Institut für Biochemie, Universität zu Koln, An der Bottmühle 2, D-5000 Koln 1, West Germany

Wee, James L.: Department of Natural Science, Loyola University, 6363 St. Charles Ave, New Orleans, LA 70118

Wetherbee, Richard: School of Botany, University of Melbourne, Parkville, Victoria Australia 3052

Yentsch, Clarice M.: Department of Flow Cytometry, Bigelow Laboratory for Ocean Sciences, McKown Pt, West Boothbay Harbor, ME 04575

Zuccarello, Giuseppe: Department of Marine Science, University of California at Santa Cruz, 214 Applied Sciences, Santa Cruz, CA 95064

Index

321

Lamoxirene, 203, 204, 206, 210
"Large vesicle", 154
Lectins, 140–141, 142–143
Lessoniaceae, 206
Linkage group XIX, in *Chlamydomonas reinhardtii*
  basal body assembly and, 42, 43–44
  behavior, 43–44
  circular gene map, 45–46
  flagellar assembly and, 40, 41–44
  herbicide-resistance and, 43, 48
  mutation interactions, 47–48
  recombination, 46–47
  related gene clustering, 41–44
Lithodesmium, 3, 7
Luciferase, 298
Luciferin-binding protein, 298

Mallomonas, scales and bristles, 95–96
Mallomonas papillosa, scales and bristles, 96–97
Mallomonas splendens
  bristle deployment, 96, 98–100
  bristle formation, 96–97, 101
Mallomonas striata, scales and bristles, 96–97
Melosira, spindle elongation, 7
Membrane
  agglutinability. *See* Agglutinin(s)
  photosynthetic. *See* Photosynthetic membrane
  recognition site, 155–156
  thylakoid. *See* Thylakoid membrane
Micrasterias, cytomorphogenesis, 149, 150, 151, 152, 153–154, 155–156, 157, 159–160, 162
  culture medium, 150
  theoretical models, 152
Microfilament
  in cytoplasmic morphogenesis
    in cyst morphogenesis, 79–81
    headed streaming bands, 73–76
    in organelle transplant, 73–75
    polymerization, 75
  in cytoplasmic movement, 55, 56–57, 58, 62
  scale/spine-associated, 93, 95–97, 102–106
    actin-based, 95–97, 102–106

centrin-based, 102, 105, 106
  in zygote polarity, 114–116
Microtubule
  in cytoplasmic morphogenesis, 81, 83–87
    in cyst morphogenesis, 77, 78, 79, 80
    headed streaming bands, 73–74
    nuclei translocation on, 75–76
  in cytoplasmic movement, 55, 61
  definition, 19–20
  flagellar, 22, 23
    basal body, 23–25, 30–31
    rootlets, 26
  of mitotic spindle
    arrangement, 5, 6, 7, 9, 15
    number, 8
    polymerization, 13–14
  scale/spine-associated, 93, 95, 96–97, 102–106
Microtubule-associated protein, 13
Microtubule-associated protein-2 (MAP-2), 76
Mitosis
  centrin dynamics during, 21, 30–31
  as research problem, 4–5
  spindle elongation during
    activation, 8
    ATP and, 10, 11, 12, 13, 14
    chromosomal movement, 4–5, 7
    microtubular polymerization, 13–14
    models, 14–15
    "motor" for, 8, 10, 13, 14–15
    phosphorylation studies, 11–13
    physiological studies, 10–11
    Stephanopyxis turris system, 4, 6, 8–15
    structural aspects, 5–8
Monoclonal antibody, 65, 178–179, 188, 192–193
Morphogenesis
  cytoskeletal regulation
    cyst morphogenesis, 75, 76, 77–81
    headed streaming bands (HSB), 73–76
    inhibitor studies, 71, 73, 81–87
    microfilaments, 73–75, 79–81
    microtubules, 73–74, 75–87
    nuclei translocation, 75–76
    primordia, 72–73
    tip growth, 72

RANDALL LIBRARY-UNCW

3 0490 0360431 X